CONSTRUCTION CRAFTS CORE UNITS

LEVEL 3 DIPLOMA

Nelson Thornes

Published in 2013 by:
Nelson Thornes Ltd
Delta Place
27 Bath Road
CHELTENHAM
GL53 7TH
United Kingdom

13 14 15 16 17 / 10 9 8 7 6 5 4 3 2 1

A catalogue record for this book is available from the British Library
ISBN 978 1 4085 2313 1

Cover photograph: Fotolia

Page make-up by GreenGate Publishing Services, Tonbridge, Kent

Printed in China by 1010 Printing International Ltd

Note to learners and tutors

This book clearly states that a risk assessment should be undertaken and the correct PPE worn for the particular activities before any practical activity is carried out. Risk assessments were carried out before photographs for this book were taken and the models are wearing the PPE deemed appropriate for the activity and situation. This was correct at the time of going to print. Colleges may prefer that their learners wear additional items of PPE not featured in the photographs in this book and should instruct learners to do so in the standard risk assessments they hold for activities undertaken by their learners. Learners should follow the standard risk assessments provided by their college for each activity they undertake which will determine the PPE they wear.

CONTENTS

Introduction   iv

Contributors to this book   v

1 Health, Safety and Welfare in
Construction and Associated Industries   **1**

2 Analysing Technical Information,
Quantities and Communication with
Others   **39**

3 Analysing the Construction Industry
and Built Environment   **77**

Index   113

Acknowledgements   114

INTRODUCTION

About this book

This book has been written to support the core units of the Cskills Awards Level 3 construction Diplomas. It covers all three mandatory units of the diploma, so you can use this book for any of the new construction crafts diplomas.

This book contains a number of features to help you acquire the knowledge you need. We've included additional features to show how the skills and knowledge can be applied to the workplace, as well as tips and advice on how you can improve your chances of gaining employment.

The features include:

KEY TERMS

DID YOU KNOW?

PRACTICAL TIP

REED TIP

CASE STUDY

TEST YOURSELF

* chapter openers which list the learning outcomes you must achieve in each unit

* key terms that provide explanations of important terminology that you will need to know and understand

* Did you know? margin notes to provide key facts that are helpful to your learning

* practical tips to explain facts or skills to remember when undertaking practical tasks

* Reed tips to offer advice about work, building your CV and how to apply the skills and knowledge you have learnt in the workplace

* case studies that are based on real tradespeople who have undertaken apprenticeships and explain why the skills and knowledge you learn with your training provider are useful in the workplace

* Test yourself multiple choice questions that appear at the end of each unit to give you the chance to revise what you have learnt and to practise your assessment (your tutor will give you the answers to these questions).

Further support for this book can be found at our website,

www.planetvocational.com/subjects/build

CONTRIBUTORS TO THIS BOOK

Reed Property & Construction

Reed Property & Construction specialises in placing staff at all levels, in both temporary and permanent positions, across the complete lifecycle of the construction process. Our consultants work with most major construction companies in the UK and our clients are involved with the design, build and maintenance of infrastructure projects throughout the UK.

Expert help

As a leading recruitment consultancy for mid–senior level construction staff in the UK, Reed Property & Construction is ideally placed to advise new workers entering the sector, from building a CV to providing expertise and sharing our extensive sector knowledge with you. That's why, throughout this book, you will find helpful hints from our highly experienced consultants, all designed to help you find that first step on the construction career ladder. These tips range from advice on CV writing to interview tips and techniques, and are all linked in with the learning material in this book.

Work-related advice

Reed Property & Construction has gained insights from some of our biggest clients – leading recruiters within the industry – to help you understand the mind-set of potential employers. This includes the traits and skills that they would like to see in their new employees, why you need the skills taught in this book and how they are used on a day to day basis within their organisations.

Getting your first job

This invaluable information is not available anywhere else and is all geared towards helping you gain a position with an employer once you've completed your studies. Entry level positions are not usually offered by recruitment companies, but the advice we've provided will help you to apply for jobs in construction and hopefully gain your first position as a skilled worker.

CONTRIBUTORS TO THIS BOOK

The case studies in this book feature staff from Laing O'Rourke and South Tyneside Homes.

Laing O'Rourke is an international engineering company that constructs large-scale building projects all over the world. Originally formed from two companies, John Laing (founded in 1848) and R O'Rourke and Son (founded in 1978) joined forces in 2001.

At Laing O'Rourke, there is a strong and unique apprenticeship programme. It runs a four-year 'Apprenticeship Plus' scheme in the UK, combining formal college education with on-the-job training. Apprentices receive support and advice from mentors and experienced tradespeople, and are given the option of three different career pathways upon completion: remaining on site, continuing into a further education programme, or progressing into supervision and management.

The company prides itself on its people development, supporting educational initiatives and investing in its employees. Laing O'Rourke believes in collaboration and teamwork as a path to achieving greater success, and strives to maintain exceptionally high standards in workplace health and safety.

South Tyneside Council's
Housing Company

South Tyneside Homes was launched in 2006, and was previously part of South Tyneside Council. It now works in partnership with the council to repair and maintain 18,000 properties within the borough, including delivering parts of the Decent Homes Programme.

South Tyneside Homes believes in putting back into the community, with 90 per cent of its employees living in the borough itself. Equality and diversity, as well as health and wellbeing of staff, is a top priority, and it has achieved the Gold Status Investors in People Award.

South Tyneside Homes is committed to the development of its employees, providing opportunities for further education and training and great career paths within the company – 80 per cent of its management team started as apprentices with the company. As well as looking after its staff and their community, the company looks after the environment too, running a renewable energy scheme for council tenants in order to reduce carbon emissions and save tenants money.

The apprenticeship programme at South Tyneside Homes has been recognised nationally, having trained over 80 young people in five main trade areas over the past six years. One of the UK's Top 100 Apprenticeship Employers, it is an Ambassador on the panel of the National Apprentice Service. It has won the Large Employer of the Year Award at the National Apprenticeship Awards and several of its apprentices have been nominated for awards, including winning the Female Apprentice of the Year for the local authority.

Chapter 1

Unit CSA–L1Core01

HEALTH, SAFETY AND WELFARE IN CONSTRUCTION AND ASSOCIATED INDUSTRIES

LEARNING OUTCOMES

LO1: Know the health and safety regulations, roles and responsibilities

LO2: Know the accident and emergency procedures and how to report them

LO3: Know how to identify hazards on construction sites

LO4: Know about health and hygiene in a construction environment

LO5: Know how to handle and store materials and equipment safely

LO6: Know about basic working platforms and access equipment

LO7: Know how to work safely around electricity in a construction environment

LO8: Know how to use personal protective equipment (PPE) correctly

LO9: Know the fire and emergency procedures

LO10: Know about signs and safety notices

INTRODUCTION

The aim of this chapter is to:

* help you to source relevant safety information
* help you to use the relevant safety procedures at work.

HEALTH AND SAFETY REGULATIONS, ROLES AND RESPONSIBILITIES

The construction industry can be dangerous, so keeping safe and healthy at work is very important. If you are not careful, you could injure yourself in an accident or perhaps use equipment or materials that could damage your health. Keeping safe and healthy will help ensure that you have a long and injury-free career.

Although the construction industry is much safer today than in the past, more than 2,000 people are injured and around 50 are killed on site every year. Many others suffer from long-term ill-health such as deafness, spinal damage, skin conditions or breathing problems.

Key health and safety legislation

Laws have been created in the UK to try to ensure safety at work. Ignoring the rules can mean injury or damage to health. It can also mean losing your job or being taken to court.

The two main laws are the Health and Safety at Work etc. Act **(HASAWA)** and the Control of Substances Hazardous to Health Regulations **(COSHH)**.

The Health and Safety at Work etc. Act (HASAWA) (1974)
This law applies to all working environments and to all types of worker, sub-contractor, employer and all visitors to the workplace. It places a duty on everyone to follow rules in order to ensure health, safety and welfare. Businesses must manage health and safety risks, for example by providing appropriate training and facilities. The Act also covers first aid, accidents and ill health.

Reporting of Injuries, Diseases and Dangerous Occurrences Regulations (RIDDOR) (1995)
Under RIDDOR, employers are required to report any injuries, diseases or dangerous occurrences to the **Health and Safety Executive (HSE)**. The regulations also state the need to maintain an **accident book**.

Control of Substances Hazardous to Health (COSHH) (2002)

In construction, it is common to be exposed to substances that could cause ill health. For example, you may use oil-based paints or preservatives, or work in conditions where there is dust or bacteria.

Employers need to protect their employees from the risks associated with using hazardous substances. This means assessing the risks and deciding on the necessary precautions to take.

Any control measures (things that are being done to reduce the risk of people being hurt or becoming ill) have to be introduced into the workplace and maintained; this includes monitoring an employee's exposure to harmful substances. The employer will need to carry out health checks and ensure that employees are made aware of the dangers and are supervised.

Control of Asbestos at Work Regulations (2012)

Asbestos was a popular building material in the past because it was a good insulator, had good fire protection properties and also protected metals against corrosion. Any building that was constructed before 2000 is likely to have some asbestos. It can be found in pipe insulation, boilers and ceiling tiles. There is also asbestos cement roof sheeting and there is a small amount of asbestos in decorative coatings such as Artex.

Asbestos has been linked with lung cancer, other damage to the lungs and breathing problems. The regulations require you and your employer to take care when dealing with asbestos:

* You should always assume that materials contain asbestos unless it is obvious that they do not.

* A record of the location and condition of asbestos should be kept.

* A risk assessment should be carried out if there is a chance that anyone will be exposed to asbestos.

The general advice is as follows:

* Do not remove the asbestos. It is not a hazard unless it is removed or damaged.

* Remember that not all asbestos presents the same risk. Asbestos cement is less dangerous than pipe insulation.

* Call in a specialist if you are uncertain.

Provision and Use of Work Equipment Regulations (PUWER) (1998)

PUWER concerns health and safety risks related to equipment used at work. It states that any risks arising from the use of equipment must either be prevented or controlled, and all suitable safety measures must have been taken. In addition, tools need to be:

* suitable for their intended use

* safe

REED TIP

Employers will want to know that you understand the importance of health and safety. Make sure you know the reasons for each safe working practice.

* well maintained

* used only by those who have been trained to do so.

Manual Handling Operations Regulations (1992)

These regulations try to control the risk of injury when lifting or handling bulky or heavy equipment and materials. The regulations state as follows:

* Hazardous manual handling should be avoided if possible.

* An assessment of hazardous manual handling should be made to try to find alternatives.

* You should use mechanical assistance where possible.

* The main idea is to look at how manual handling is carried out and finding safer ways of doing it.

Personal Protection at Work Regulations (PPE) (1992)

This law states that employers must provide employees with personal protective equipment **(PPE)** at work whenever there is a risk to health and safety. PPE needs to be:

* suitable for the work being done

* well maintained and replaced if damaged

* properly stored

* correctly used (which means employees need to be trained in how to use the PPE properly).

Work at Height Regulations (2005)

Whenever a person works at any height there is a risk that they could fall and injure themselves. The regulations place a duty on employers or anyone who controls the work of others. This means that they need to:

* plan and organise the work

* make sure those working at height are **competent**

* assess the risks and provide appropriate equipment

* manage work near or on fragile surfaces

* ensure equipment is inspected and maintained.

In all cases the regulations suggest that, if it is possible, work at height should be avoided. Perhaps the job could be done from ground level? If it is not possible, then equipment and other measures are needed to prevent the risk of falling. When working at height measures also need to be put in place to minimise the distance someone might fall.

KEY TERMS

PPE

– personal protective equipment can include gloves, goggles and hard hats.

Competent

– to be competent an organisation or individual must have:

* sufficient knowledge of the tasks to be undertaken and the risks involved

* the experience and ability to carry out their duties in relation to the project, to recognise their limitations and take appropriate action to prevent harm to those carrying out construction work, or those affected by the work.

(*Source* HSE)

Figure 1.1 Examples of personal protective equipment

Employer responsibilities under HASAWA

HASAWA states that employers with five or more staff need their own health and safety policy. Employers must assess any risks that may be involved in their workplace and then introduce controls to reduce these risks. These risk assessments need to be reviewed regularly.

Employers also need to supply personal protective equipment (PPE) to all employees when it is needed and to ensure that it is worn when required.

Specific employer responsibilities are outlined in Table 1.1.

Employee responsibilities under HASAWA

HASAWA states that all those operating in the workplace must aim to work in a safe way. For example, they must wear any PPE provided and look after their equipment. Employees should not be charged for PPE or any actions that the employer needs to take to ensure safety.

Specific employer responsibilities are outlined in Table 1.1. Table 1.2 identifies the key employee responsibilities.

KEY TERMS

Risk

– the likelihood that a person may be harmed if they are exposed to a hazard.

Hazard

– a potential source of harm, injury or ill-health.

Near miss

– any incident, accident or emergency that did not result in an injury but could have done so.

Employer responsibility	Explanation
Safe working environment	Where possible all potential **risks** and **hazards** should be eliminated.
Adequate staff training	When new employees begin a job their induction should cover health and safety. There should be ongoing training for existing employees on risks and control measures.
Health and safety information	Relevant information related to health and safety should be available for employees to read and have their own copies.
Risk assessment	Each task or job should be investigated and potential risks identified so that measures can be put in place. A risk assessment and method statement should be produced. The method statement will tell you how to carry out the task, what PPE to wear, equipment to use and the sequence of its use.
Supervision	A competent and experienced individual should always be available to help ensure that health and safety problems are avoided.

Table 1.1 Employer responsibilities under HASAWA

Employee responsibility	Explanation
Working safely	Employees should take care of themselves, only do work that they are competent to carry out and remove obvious hazards if they are seen.
Working in partnership with the employer	Co-operation is important and you should never interfere with or misuse any health and safety signs or equipment. You should always follow the site rules.
Reporting hazards, **near misses** and accidents correctly	Any health and safety problems should be reported and discussed, particularly a near miss or an actual accident.

Table 1.2 Employee responsibilities under HASAWA

Health and Safety Executive

The Health and Safety Executive (HSE) is responsible for health, safety and welfare. It carries out spot checks on different workplaces to make sure that the law is being followed.

HSE inspectors have access to all areas of a construction site and can also bring in the police. If they find a problem then they can issue an **improvement notice**. This gives the employer a limited amount of time to put things right.

In serious cases, the HSE can issue a **prohibition notice**. This means all work has to stop until the problem is dealt with. An employer, the employees or **sub-contractors** could be taken to court.

The roles and responsibilities of the HSE are outlined in Table 1.3.

Responsibility	Explanation
Enforcement	It is the HSE's responsibility to reduce work-related death, injury and ill health. It will use the law against those who put others at risk.
Legislation and advice	The HSE will use health and safety legislation to serve improvement or prohibition notices or even to prosecute those who break health and safety rules. Inspectors will provide advice either face-to-face or in writing on health and safety matters.
Inspection	The HSE will look at site conditions, standards and practices and inspect documents to make sure that businesses and individuals are complying with health and safety law.

Table 1.3 HSE roles and responsibilities

Sources of health and safety information

There is a wide variety of health and safety information. Most of it is available free of charge, while other organisations may make a charge to provide information and advice. Table 1.4 outlines the key sources of health and safety information.

Source	Types of information	Website
Health and Safety Executive (HSE)	The HSE is the primary source of work-related health and safety information. It covers all possible topics and industries.	www.hse.gov.uk
Construction Industry Training Board (CITB)	The national training organisation provides key information on legislation and site safety.	www.citb.co.uk
British Standards Institute (BSI)	Provides guidelines for risk management, PPE, fire hazards and many other health and safety-related areas.	www.bsigroup.com
Royal Society for the Prevention of Accidents (RoSPA)	Provides training, consultancy and advice on a wide range of health and safety issues that are aimed to reduce work related accidents and ill health.	www.rospa.com
Royal Society for Public Health (RSPH)	Has a range of qualifications and training programmes focusing on health and safety.	www.rsph.org.uk

Table 1.4 Health and safety information

Informing the HSE

The HSE requires the reporting of:

* deaths and injuries – any **major injury**, **over 7-day injury** or death

* occupational disease

* dangerous occurrence – a collapse, explosion, fire or collision

* gas accidents – any accidental leaks or other incident related to gas.

Enforcing guidance

Work-related injuries and illnesses affect huge numbers of people. According to the HSE, 1.1 million working people in the UK suffered from a work-related illness in 2011 to 2012. Across all industries, 173 workers were killed, 111,000 other injuries were reported and 27 million working days were lost.

The construction industry is a high risk one and, although only around 5 per cent of the working population is in construction, it accounts for 10 per cent of all major injuries and 22 per cent of fatal injuries.

The good news is that enforcing guidance on health and safety has driven down the numbers of injuries and deaths in the industry. Only 20 years ago over 120 construction workers died in workplace accidents each year. This is now reduced to fewer than 60 a year.

However, there is still more work to be done and it is vital that organisations such as the HSE continue to enforce health and safety and continue to reduce risks in the industry.

On-site safety inductions and toolbox talks

The HSE suggests that all new workers arriving on site should attend a short induction session on health and safety. It should:

* show the commitment of the company to health and safety

* explain the health and safety policy

* explain the roles individuals play in the policy

* state that each individual has a legal duty to contribute to safe working

* cover issues like excavations, work at height, electricity and fire risk

* provide a layout of the site and show evacuation routes

* identify where fire fighting equipment is located

* ensure that all employees have evidence of their skills

* stress the importance of signing in and out of the site.

KEY TERMS

Major injury

– any fractures, amputations, dislocations, loss of sight or other severe injury.

Over 7-day injury

– an injury that has kept someone off work for more than seven days.

DID YOU KNOW?

Workplace injuries cost the UK £13.4bn in 2010 to 2011.

DID YOU KNOW?

Toolbox talks are normally given by a supervisor and often take place on site, either during the course of a normal working day or when someone has been seen working in an unsafe way. CITB produces a book called *GT700 Toolbox Talks* which covers a range of health and safety topics, from trying a new process and using new equipment to particular hazards or work practices.

PRACTICAL TIP

If you come across any health and safety problems you should report them so that they can be controlled.

Figure 1.2 It's important that you know where your company's fire-fighting equipment is located

Behaviour and actions that could affect others

It is the responsibility of everyone on site not only to look after their own health and safety, but also to ensure that their actions do not put anyone else at risk.

Trying to carry out work that you are not competent to do is not only dangerous to yourself but could compromise the safety of others.

Simple actions, such as ensuring that all of your rubbish and waste is properly disposed of, will go a long way to removing hazards on site that could affect others.

Just as you should not create a hazard, ignoring an obvious one is just as dangerous. You should always obey site rules and particularly the health and safety rules. You should follow any instructions you are given.

ACCIDENT AND EMERGENCY PROCEDURES

All sites will have specific procedures for dealing with accidents and emergencies. An emergency will often mean that the site needs to be evacuated, so you should know in advance where to assemble and who to report to. The site should never be re-entered without authorisation from an individual in charge or the emergency services.

Types of emergencies

Emergencies are incidents that require immediate action. They can include:

* fires
* spillages or leaks of chemicals or other hazardous substances, such as gas
* failure of a scaffold
* collapse of a wall or trench
* a health problem
* an injury
* bombs and security alerts.

Legislation and reporting accidents

RIDDOR (1995) puts a duty on employers, anyone who is self-employed, or an individual in control of the work, to report any serious workplace accidents, occupational diseases or dangerous occurrences (also known as near misses).

The report has to be made by these individuals and, if it is serious enough, the responsible person may have to fill out a RIDDOR report.

Injuries, diseases and dangerous occurrences

Construction sites can be dangerous places, as we have seen. The HSE maintains a list of all possible injuries, diseases and dangerous occurrences, particularly those that need to be reported.

Injuries

There are two main classifications of injuries: minor and major. A minor injury can usually be handled by a competent first aider, although it is often a good idea to refer the individual to their doctor or to the hospital. Typical minor injuries can include:

* minor cuts
* minor burns
* exposure to fumes.

Major injuries are more dangerous and will usually require the presence of an ambulance with paramedics. Major injuries can include:

* bone fracture
* concussion
* unconsciousness
* electric shock.

Diseases

There are several different diseases and health issues that have to be reported, particularly if a doctor notifies that a disease has been diagnosed. These include:

* poisoning
* infections
* skin diseases
* occupational cancer
* lung diseases
* hand/arm vibration syndrome.

Dangerous occurrences

Even if something happens that does not result in an injury, but could easily have done so, it is classed as a dangerous occurrence. It needs to be reported immediately and then followed up by an accident report form. Dangerous occurrences can include:

* accidental release of a substance that could damage health

* anything coming into contact with overhead power lines

* an electrical problem that caused a fire or explosion

* collapse or partial collapse of scaffolding over 5 m high.

PRACTICAL TIP

An up-to-date list of dangerous occurrences is maintained by the Health and Safety Executive.

Recording accidents and emergencies

The Reporting of Injuries, Diseases and Dangerous Occurrences Regulations (RIDDOR) (1995) requires employers to:

* report any relevant injuries, diseases or dangerous occurrences to the Health and Safety Executive (HSE)

* keep records of incidents in a formal and organised manner (for example, in an accident book or online database).

After an accident, you may need to complete an accident report form – either in writing or online. This form may be completed by the person who was injured or the first aider.

On the accident report form you need to note down:

* the casualty's personal details, e.g. name, address, occupation
* the name of the person filling in the report form
* the details of the accident.

In addition, the person reporting the accident will need to sign the form.

On site a trained first aider will be the first individual to try and deal with the situation. In addition to trying to save life, stop the condition from getting worse and getting help, they will also record the occurrence.

On larger sites there will be a health and safety officer, who would keep records and documentation detailing any accidents and emergencies that have taken place on site. All companies should keep such records; it may be a legal requirement for them to do so under RIDDOR and it is good practice to do so in case the HSE asks to see it.

Importance of reporting accidents and near misses

Reporting incidents is not just about complying with the law or providing information for statistics. Each time an accident or near miss takes place it means lessons can be learned and future problems avoided.

The accident or near miss can alert the business or organisation to a potential problem. They can then take steps to ensure that it does not occur in the future.

Major and minor injuries and near misses

RIDDOR defines a major injury as:

* a fracture (but not to a finger, thumb or toes)
* a dislocation
* an amputation
* a loss of sight in an eye
* a chemical or hot metal burn to the eye
* a penetrating injury to the eye
* an electric shock or electric burn leading to unconsciousness and/or requiring resuscitation
* hyperthermia, heat-induced illness or unconsciousness
* asphyxia
* exposure to a harmful substance
* inhalation of a substance
* acute illness after exposure to toxins or infected materials.

A minor injury could be considered as any occurrence that does not fall into any of the above categories.

A near miss is any incident that did not actually result in an injury but which could have caused a major injury if it had done so. Non-reportable near misses are useful to record as they can help to identify potential problems. Looking at a list of near misses might show patterns for potential risk.

Accident trends

We have already seen that the HSE maintains statistics on the number and types of construction accidents. The following are among the 2011/2012 construction statistics:

* There were 49 fatalities.

* There were 5,000 occupational cancer patients.

* There were 74,000 cases of work-related ill health.

* The most common types of injury were caused by falls, although many injuries were caused by falling objects, collapses and electricity. A number of construction workers were also hurt when they slipped or tripped, or were injured while lifting heavy objects.

Accidents, emergencies and the employer

Even less serious accidents and injuries can cost a business a great deal of money. But there are other costs too:

* Poor company image – if a business does not have health and safety controls in place then it may get a reputation for not caring about its employees. The number of accidents and injuries may be far higher than average.

* Loss of production – the injured individual might have to be treated and then may need a period of time off work to recover. The loss of production can include those who have to take time out from working to help the injured person and the time of a manager or supervisor who has to deal with all the paperwork and problems.

* Insurance – each time there is an accident or injury claim against the company's insurance the premiums will go up. If there are many accidents and injuries the business may find it impossible to get insurance. It is a legal requirement for a business to have insurance so in the end that company might have to close down.

* Closure of site – if there is a serious accident or injury then the site may have to be closed while investigations take place to discover the reason, or who was responsible. This could cause serious delays and loss of income for workers and the business.

DID YOU KNOW?

RoSPA (the Royal Society for the Prevention of Accidents) uses many of the statistics from the HSE. The latest figures that RoSPA has analysed date back to 2008/2009. In that year, 1.2 million people in the UK were suffering from work-related illnesses. With fewer than 132,000 reportable injuries at work, this is believed to be around half of the real figure.

DID YOU KNOW?

An employee working in a small business broke two bones in his arm. He could not return to proper duties for eight months. He lost out on wages while he was off sick and, in total, it cost the business over £45,000.

REED TIP

On some construction sites, you may get a Health and Safety Inspector come to look round without any notice – one more reason to always be thinking about working safely.

Accident and emergency authorised personnel

Several different groups of people could be involved in dealing with accident and emergency situations. These are listed in Table 1.5.

Authorised personnel	Role
First aiders and emergency responders	These are employees on site and in the workforce who have been trained to be the first to respond to accidents and injuries. The minimum provision of an appointed person would be someone who has had basic first aid training. The appointment of a first aider is someone who has attained a higher or specific level of training. A construction site with fewer than 5 employees needs an appointed first aider. A construction site with up to 50 employees requires a trained first aider, and for bigger sites at least one trained first aider is required for every 50 people.
Supervisors and managers	These have the responsibility of managing the site and would have to organise the response and contact emergency services if necessary. They would also ensure that records of any accidents are completed and up to date and notify the HSE if required.
Health and Safety Executive	The HSE requires businesses to investigate all accidents and emergencies. The HSE may send an inspector, or even a team, to investigate and take action if the law has been broken.
Emergency services	Calling the emergency services depends on the seriousness of the accident. Paramedics will take charge of the situation if there is a serious injury and if they feel it necessary will take the individual to hospital.

Table 1.5 People who deal with accident and emergency situations

DID YOU KNOW?

The three main emergency services in the UK are: the Fire Service (for fire and rescue); the Ambulance Service (for medical emergencies); the Police (for an immediate police response). Call them on 999 only if it is an emergency.

The basic first aid kit

BS 8599 relates to first aid kits, but it is not legally binding. The contents of a first aid box will depend on an employer's assessment of their likely needs. The HSE does not have to approve the contents of a first aid box but it states that where the work involves low level hazards the minimum contents of a first aid box should be:

* a copy of its leaflet on first aid – *HSE Basic advice on first aid at work*

* 20 sterile plasters of assorted size

* 2 sterile eye pads

* 4 sterile triangular bandages

* 6 safety pins

* 2 large sterile, unmedicated wound dressings

* 6 medium-sized sterile unmedicated wound dressings

* 1 pair of disposable gloves.

The HSE also recommends that no tablets or medicines are kept in the first aid box.

Figure 1.3 A typical first aid box

What to do if you discover an accident

When an accident happens it may not only injure the person involved directly, but it may also create a hazard that could then injure others. You need to make sure that the area is safe enough for you or someone else to help the injured person. It may be necessary to turn off the electrical supply or remove obstructions to the site of the accident.

The first thing that needs to be done if there is an accident is to raise the alarm. This could mean:

* calling for the first aider

* phoning for the emergency services

* dealing with the problem yourself.

How you respond will depend on the severity of the injury.

You should follow this procedure if you need to contact the emergency services:

* Find a telephone away from the emergency.

* Dial 999.

* You may have to go through a switchboard. Carefully listen to what the operator is saying to you and try to stay calm.

* When asked, give the operator your name and location, and the name of the emergency service or services you require.

* You will then be transferred to the appropriate emergency service, who will ask you questions about the accident and its location. Answer the questions in a clear and calm way.

* Once the call is over, make sure someone is available to help direct the emergency services to the location of the accident.

IDENTIFYING HAZARDS

As we have already seen, construction sites are potentially dangerous places. The most effective way of handling health and safety on a construction site is to spot the hazards and deal with them before they can cause an accident or an injury. This begins with basic housekeeping and carrying out risk assessments. It also means having a procedure in place to report hazards so that they can be dealt with.

Good housekeeping

Work areas should always be clean and tidy. Sites that are messy, strewn with materials, equipment, wires and other hazards can prove to be very dangerous. You should:

* always work in a tidy way

* never block fire exits or emergency escape routes

* never leave nails and screws scattered around

* ensure you clean and sweep up at the end of each working day

* not block walkways

* never overfill skips or bins

* never leave food waste on site.

Risk assessments and method statements

It is a legal requirement for employers to carry out risk assessments. This covers not only those who are actually working on a particular job, but other workers in the immediate area, and others who might be affected by the work.

It is important to remember that when you are carrying out work your actions may affect the safety of other people. It is important, therefore, to know whether there are any potential hazards. Once you know what these hazards are you can do something to either prevent or reduce them as a risk. Every job has potential hazards.

There are five simple steps to carrying out a risk assessment, which are shown in Table 1.6, using the example of repointing brickwork on the front face of a dwelling.

Step	Action	Example
1	Identify hazards	The property is on a street with a narrow pavement. The damaged brickwork and loose mortar need to be removed and placed in a skip below. Scaffolding has been erected. The road is not closed to traffic.
2	Identify who is at risk	The workers repointing are at risk as they are working at height. Pedestrians and vehicles passing are at risk from the positioning of the skip and the chance that debris could fall from height.
3	What is the risk from the hazard that may cause an accident?	The risk to the workers is relatively low as they have PPE and the scaffolding has been correctly erected. The risk to those passing by is higher, as they are unaware of the work being carried out above them.
4	Measures to be taken to reduce the risk	Station someone near the skip to direct pedestrians and vehicles away from the skip while the work is being carried out. Fix a secure barrier to the edge of the scaffolding to reduce the chance of debris falling down. Lower the bricks and mortar debris using a bucket or bag into the skip and not throwing them from the scaffolding. Consider carrying out the work when there are fewer pedestrians and less traffic on the road.
5	Monitor the risk	If there are problems with the first stages of the job, you need to take steps to solve them. If necessary consider taking the debris by hand through the building after removal.

Table 1.6 A five-step risk assessment for repointing brickwork

Your employer should follow these working practices, which can help to prevent accidents or dangerous situations occurring in the workplace:

* *Risk assessments* look carefully at what could cause an individual harm and how to prevent this. This is to ensure that no one should be injured or become ill as a result of their work. Risk assessments identify how likely it is that an accident might happen and the consequences of it happening. A risk factor is worked out and control measures created to try to offset them.

* *Method statements,* however brief, should be available for every risk assessment. They summarise risk assessments and other findings to provide guidance on how the work should be carried out.

* *Permit to work systems* are used for very high risk or even potentially fatal activities. They are checklists that need to be completed before the work begins. They must be signed by a supervisor.

* *A hazard book* lists standard tasks and identifies common hazards. These are useful tools to help quickly identify hazards related to particular tasks.

Types of hazards

Typical construction accidents can include:

* fires and explosions

* slips, trips and falls

* burns, including those from chemicals

* falls from scaffolding, ladders and roofs

* electrocution

* injury from faulty machinery

* power tool accidents

* being hit by construction debris

* falling through holes in flooring.

We will look at some of the more common hazards in a little more detail.

Fires

Fires need oxygen, heat and fuel to burn. Even a spark can provide enough heat needed to start a fire, and anything flammable, such as petrol, paper or wood, provides the fuel. It may help to remember the 'triangle of fire' – heat, oxygen and fuel are all needed to make fire so remove one or more to help prevent or stop the fire.

Tripping

Leaving equipment and materials lying around can cause accidents, as can trailing cables and spilt water or oil. Some of these materials are also potential fire hazards.

Chemical spills

If the chemicals are not hazardous then they just need to be mopped up. But sometimes they do involve hazardous materials and there will be an existing plan on how to deal with them. A risk assessment will have been carried out.

Falls from height

A fall even from a low height can cause serious injuries. Precautions need to be taken when working at height to avoid permanent injury. You should also consider falls into open excavations as falls from height. All the same precautions need to be in place to prevent a fall.

Burns

Burns can be caused not only by fires and heat, but also from chemicals and solvents. Electricity and wet concrete and cement can also burn skin. PPE is often the best way to avoid these dangers. Sunburn is a common and uncomfortable form of burning and sunscreen should be made available. For example, keeping skin covered up will help to prevent sunburn. You might think a tan looks good, but it could lead to skin cancer.

Electrical

Electricity is hazardous and electric shocks can cause burns and muscle damage, and can kill.

Exposure to hazardous substances

We look at hazardous substances in more detail on pages 20–1. COSHH regulations identify hazardous substances and require them to be labelled. You should always follow the instructions when using them.

Plant and vehicles

On busy sites there is always a danger from moving vehicles and heavy plant. Although many are fitted with reversing alarms, it may not be easy to hear them over other machinery and equipment. You should always ensure you are not blocking routes or exits. Designated walkways separate site traffic and pedestrians – this includes workers who are walking around the site. Crossing points should be in place for ease of movement on site.

Reporting hazards

We have already seen that hazards have the potential to cause serious accidents and injuries. It is therefore important to report hazards and there are different methods of doing this.

The first major reason to report hazards is to prevent danger to others, whether they are other employees or visitors to the site. It is vital to prevent accidents from taking place and to quickly correct any dangerous situations.

Injuries, diseases and actual accidents all need to be reported and so do dangerous occurrences. These are incidents that do not result in an actual injury, but could easily have hurt someone.

Accidents need to be recorded in an accident book, computer database or other secure recording system, as do near misses. Again it is a legal requirement to keep appropriate records of accidents and every company will have a procedure for this which they should tell you about. Everyone should know where the book is kept or how the records are made. Anyone that has been hurt or has taken part in dealing with an occurrence should complete the details of what has happened. Typically this will require you to fill in:

* the date, time and place of the incident

* how it happened

* what was the cause

* how it was dealt with

* who was involved

* signature and date.

The details in the book have to be transferred onto an official HSE report form.

As far as is possible, the site, company or workplace will have set procedures in place for reporting hazards and accidents. These procedures will usually be found in the place where the accident book or records are stored. The location tends to be posted on the site notice board.

How hazards are created

Construction sites are busy places. There are constantly new stages in development. As each stage is begun a whole new set of potential hazards need to be considered.

At the same time, new workers will always be joining the site. It is mandatory for them to be given health and safety instruction during induction. But sometimes this is impossible due to pressure of work or availability of trainers.

Construction sites can become even more hazardous in times of extreme weather:

* Flooding – long periods of rain can cause trenches to fill with water, cellars to be flooded and smooth surfaces to become extremely wet and slippery.

* Wind – strong winds may prevent all work at height. Scaffolding may have become unstable, unsecured roofing materials may come loose, dry-stored materials such as sand and cement may have been blown across the site.

* Heat – this can change the behaviour of materials: setting quicker, failing to cure and melting. It can also seriously affect the health of the workforce through dehydration and heat exhaustion.

* Snow – this can add enormous weight to roofs and other structures and could cause collapse. Snow can also prevent access or block exits and can mean that simple and routine work becomes impossible due to frozen conditions.

Storing combustibles and chemicals

A combustible substance can be both flammable and explosive. There are some basic suggestions from the HSE about storing these:

* Ventilation – the area should be well ventilated to disperse any vapours that could trigger off an explosion.

* Ignition – an ignition is any spark or flame that could trigger off the vapours, so materials should be stored away from any area that uses electrical equipment or any tool that heats up.

* Containment – the materials should always be kept in proper containers with lids and there should be spillage trays to prevent any leak seeping into other parts of the site.

* Exchange – in many cases it can be possible to find an alternative material that is less dangerous. This option should be taken if possible.

* Separation – always keep flammable substances away from general work areas. If possible they should be partitioned off.

Combustible materials can include a large number of commonly used substances, such as cleaning agents, paints and adhesives.

HEALTH AND HYGIENE

Just as hazards can be a major problem on site, other less obvious problems relating to health and hygiene can also be an issue. It is both your responsibility and that of your employer to make sure that you stay healthy.

The employer will need to provide basic welfare facilities, no matter where you are working and these must have minimum standards.

Welfare facilities

Welfare facilities can include a wide range of different considerations, as can be seen in Table 1.7.

Facilities	Purpose and minimum standards
Toilets	If there is a lock on the door there is no need to have separate male and female toilets. There should be enough for the site workforce. If there is no flushing water on site they must be chemical toilets.
Washing facilities	There should be a wash basin large enough to be able to wash up to the elbow. There should be soap, hot and cold water and, if you are working with dangerous substances, then showers are needed.
Drinking water	Clean drinking water should be available; either directly connected to the mains or bottled water. Employers must ensure that there is no **contamination.**
Dry room	This can operate also as a store room, which needs to be secure so that workers can leave their belongings there and also use it as a place to dry out if they have been working in wet weather, in which case a heater needs to be provided.
Work break area	This is a shelter out of the wind and rain, with a kettle, a microwave, tables and chairs. It should also have heating.

Table 1.7 Welfare facilities in the workplace

CASE STUDY

South Tyneside Council's Housing Company

Staying safe on site

Johnny McErlane finished his apprenticeship at South Tyneside Homes a year ago.

'I've been working on sheltered accommodation for the last year, so there are a lot of vulnerable and elderly people around. All the things I learnt at college from doing the health and safety exams comes into practice really, like taking care when using extension leads, wearing high-vis and correct footwear. It's not just about your health and safety, but looking out for others as well.

On the shelters, you can get a health and safety inspector who just comes around randomly, so you have to always be ready. It just becomes a habit once it's been drilled into you. You're health and safety conscious all the time.

The shelters also have a fire alarm drill every second Monday, so you've got to know the procedure involved there. When it comes to the more specialised skills, such as mouth-to-mouth and CPR, you might have a designated first aider on site who will have their skills refreshed regularly. Having a full first aid certificate would be valuable if you're working in construction.

You cover quite a bit of the first aid skills in college and you really have to know them because you're not always working on large sites. For example, you might be on the repairs team, working in people's houses where you wouldn't have a first aider, so you've got to have the basic knowledge yourself, just in case. All our vans have a basic first aid kit that's kept fully stocked.

The company keeps our knowledge current with these "toolbox talks", which are like refresher courses. They give you any new information that needs to be passed on to all the trades. It's a good way of keeping everyone up to date.'

Noise

Ear defenders are the best precaution to protect the ears from loud noises on site. Ear defenders are either basic ear plugs or ear muffs, which can be seen in Fig 1.13 on page 32.

The long-term impact of noise depends on the intensity and duration of the noise. Basically, the louder and longer the noise exposure, the more damage is caused. There are ways of dealing with this:

* Remove the source of the noise.

* Move the equipment away from those not directly working with it.

* Put the source of the noise into a soundproof area or cover it with soundproof material.

* Ask a supervisor if they can move all other employees away from that part of the site until the noise stops.

Substances hazardous to health

COSHH Regulations (see page 3) identify a wide variety of substances and materials that must be labelled in different ways.

Controlling the use of these substances is always difficult. Ideally, their use should be eliminated (stopped) or they should be replaced with something less harmful. Failing this, they should only be used in controlled or restricted areas. If none of this is possible then they should only be used in controlled situations.

If a hazardous situation occurs at work, then you should:

* ensure the area is made safe

* inform the supervisor, site manager, safety officer or other nominated person.

You will also need to report any potential hazards or near misses.

Personal hygiene

Construction sites can be dirty places to work. Some jobs will expose you to dust, chemicals or substances that can make contact with your skin or may stain your work clothing. It is good practice to wear suitable PPE as a first line of defence as chemicals can penetrate your skin. Whenever you have finished a job you should always wash your hands. This is certainly true before eating lunch or travelling home. It can be good practice to have dedicated work clothing, which should be washed regularly.

Always ensure you wash your hands and face and scrub your nails. This will prevent dirt, chemicals and other substances from contaminating your food and your home.

Make sure that you regularly wash your work clothing and either repair it or replace it if it becomes too worn or stained.

Health risks

The construction industry uses a wide variety of substances that could harm your health. You will also be carrying out work that could be a health risk to you, and you should always be aware that certain activities could cause long-term damage or even kill you if things go wrong. Unfortunately not all health risks are immediately obvious. It is important to make sure that from time to time you have health checks, particularly if you have been using hazardous substances. Table 1.8 outlines some potential health risks in a typical construction site.

KEY TERMS

Dermatitis

– this is an inflammation of the skin. The skin will become red and sore, particularly if you scratch the area. A GP should be consulted.

Leptospirosis

– this is also known as Weil's disease. It is spread by touching soil or water contaminated with the urine of wild animals infected with the leptospira bacteria. Symptoms are usually flu-like but in extreme cases it can cause organ failure.

Health risk	Potential future problems
Dust	The most dangerous potential dust is, of course, asbestos, which **should only be handled by specialists under controlled conditions**. But even brick dust and other fine particles can cause eye injuries, problems with breathing and even cancer.
Chemicals	Inhaling or swallowing dangerous chemicals could cause immediate, long-term damage to lungs and other internal organs. Skin problems include burns or skin can become very inflamed and sore. This is known as **dermatitis**.
Bacteria	Contact with waste water or soil could lead to a bacterial infection. The germs in the water or dirt could cause infection which will require treatment if they enter the body. The most extreme version is **leptospirosis**.
Heavy objects	Lifting heavy, bulky or awkward objects can lead to permanent back injuries that could require surgery. Heavy objects can also damage the muscles in all areas of the body.
Noise	Failure to wear ear defenders when you are exposed to loud noises can permanently affect your hearing. This could lead to deafness in the future.
Vibrating tools	Using machines that vibrate can cause a condition known as hand/arm vibration syndrome (HAVS) or vibration white finger, which is caused by injury to nerves and blood vessels. You will feel tingling that could lead to permanent numbness in the fingers and hands, as well as muscle weakness.
Cuts	Any open wound, no matter how small, leaves your body exposed to potential infections. Cuts should always be cleaned and covered, preferably with a waterproof dressing. The blood loss from deep cuts could make you feel faint and weak, which may be dangerous if you are working at height or operating machinery.
Sunlight	Most construction work involves working outside. There is a temptation to take advantage of hot weather and get a tan. But long-term exposure to sunshine means risking skin cancer so you should cover up and apply sun cream.
Head injuries	You should seek medical attention after any bump to the head. Severe head injuries could cause epilepsy, hearing problems, brain damage or death.

Table 1.8 Health risks in construction

HANDLING AND STORING MATERIALS AND EQUIPMENT

On a busy construction site it is often tempting not to even think about the potential dangers of handling equipment and materials. If something needs to be moved or collected you will just pick it up without any thought. It is also tempting just to drop your tools and other equipment when you have finished with them to deal with later. But abandoned equipment and tools can cause hazards both for you and for other people.

Safe lifting

Lifting or handling heavy or bulky items is a major cause of injuries on construction sites. So whenever you are dealing with a heavy load, it is important to carry out a basic risk assessment.

The first thing you need to do is to think about the job to be done and ask:

* Do I need to lift it manually or is there another way of getting the object to where I need it?

Consider any mechanical methods of transporting loads or picking up materials. If there really is no alternative, then ask yourself:

1. Do I need to bend or twist?
2. Does the object need to be lifted or put down from high up?
3. Does the object need to be carried a long way?
4. Does the object need to be pushed or pulled for a long distance?
5. Is the object likely to shift around while it is being moved?

If the answer to any of these questions is 'yes', you may need to adjust the way the task is done to make it safer.

Think about the object itself. Ask:

1. Is it just heavy or is it also bulky and an awkward shape?
2. How easy is it to get a good hand-hold on the object?
3. Is the object a single item or are there parts that might move around and shift the weight?
4. Is the object hot or does it have sharp edges?

Again, if you have answered 'yes' to any of these questions, then you need to take steps to address these issues.

It is also important to think about the working environment and where the lifting and carrying is taking place. Ask yourself:

1. Are the floors stable?

2. Are the surfaces slippery?

3. Will a lack of space restrict my movement?

4. Are there any steps or slopes?

5. What is the lighting like?

Before lifting and moving an object, think about the following:

● Check that your pathway is clear to where the load needs to be taken.

● Look at the product data sheet and assess the weight. If you think the object is too heavy or difficult to move then ask someone to help you. Alternatively, you may need to use a mechanical lifting device.

When you are ready to lift, gently raise the load. Take care to ensure the correct posture – you should have a straight back, with your elbows tucked in, your knees bent and your feet slightly apart.

Once you have picked up the load, move slowly towards your destination. When you get there, make sure that you do not drop the load but carefully place it down.

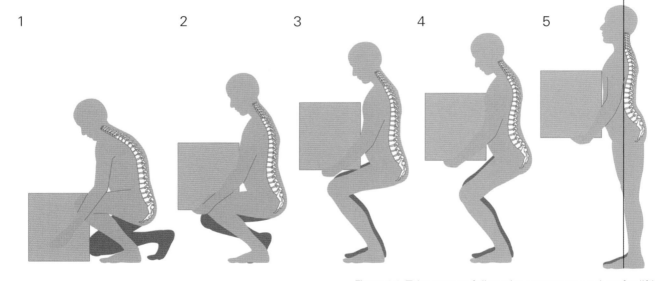

1   2   3   4   5

Figure 1.4 Take care to follow the correct procedure for lifting

Sack trolleys are useful for moving heavy and bulky items around. Gently slide the bottom of the sack trolley under the object and then raise the trolley to an angle of 45° before moving off. Make sure that the object is properly balanced and is not too big for the trolley.

Trailers and forklift trucks are often used on large construction sites, as are dump trucks. Never use these without proper training.

Figure 1.5 Pallet truck

Figure 1.6 Sack trolley

Site safety equipment

You should always read the construction site safety rules and when required wear your PPE. Simple things, such as wearing the right footwear for the right job, are important.

Safety equipment falls into two main categories:

* PPE – including hard hats, footwear, gloves, glasses and safety vests

* perimeter safety – this includes screens, netting and guards or clamps to prevent materials from falling or spreading.

Construction safety is also directed by signs, which will highlight potential hazards.

Safe handling of materials and equipment

All tools and equipment are potentially dangerous. It is up to you to make sure that they do not cause harm to yourself or others. You should always know how to use tools and equipment. This means either instruction from someone else who is experienced, or at least reading the manufacturer's instructions.

You should always make sure that you:

* use the right tool – don't be tempted to use a tool that is close to hand instead of the one that is right for the job

* wear your PPE – the one time you decide not to bother could be the time that you injure yourself

* never try to use a tool or a piece of equipment that you have not been trained to use.

You should always remember that if you are working on a building that was constructed before 2000 it may contain asbestos.

Correct storage

We have already seen that tools and equipment need to be treated with respect. Damaged tools and equipment are not only less effective at doing their job, they could also cause you to injure yourself.

Table 1.9 provides some pointers on how to store and handle different types of materials and equipment.

Materials and equipment	Safe storage and handling
Hand tools	Store hand tools with sharp edges either in a cover or a roll. They should be stored in bags or boxes. They should always be dried before putting them away as they will rust.
Power tools	Never carry them by the cable. Store them in their original carrying case. Always follow the manufacturer's instructions.
Wheelbarrows	Check the tyres and metal stays regularly. Always clean out after use and never overload.
Bricks and blocks	Never store more than two packs high. When cutting open a pack, be careful as the bricks could collapse.
Slabs and curbs	Store slabs flat on their edges on level ground, preferably with wood underneath to prevent damage. Store curbs the same way. To prevent weather damage, cover them with a sheet.
Tiles	Always cover them and protect them from damage as they are relatively fragile. Ideally store them in a hut or container.
Aggregates	Never store aggregates under trees as leaves will drop on them and contaminate them. Cover them with plastic sheets.
Plaster and plasterboard	Plaster needs to be kept dry, so even if stored inside you should take the precaution of putting the bags on pallets. To prevent moisture do not store against walls and do not pile higher than five bags. Plasterboard can be awkward to manage and move around. It also needs to be stored in a waterproof area. It should be stored flat and off the ground but should not be stored against walls as it may bend. Use a rotation system so that the materials are not stored in the same place for long periods.
Wood	Always keep wood in dry, well-ventilated conditions. If it needs to be stored outside it should be stored on bearers that may be on concrete. If wood gets wet and bends it is virtually useless. Always be careful when moving large cuts of wood or sheets of ply or MDF as they can easily become damaged.
Adhesives and paint	Always read the manufacturer's instructions. Ideally they should always be stored on clearly marked shelves. Make sure you rotate the stock using the older stock first. Always make sure that containers are tightly sealed. Storage areas must comply with fire regulations and display signs to advise of their contents.

Table 1.9 Safe storing and handling of materials and equipment

Waste control

The expectation within the building services industry is increasingly that working practices conserve energy and protect the environment. Everyone can play a part in this. For example, you can contribute by turning off hose pipes when you have finished using water, or not running electrical items when you don't need to.

Simple things, such as keeping construction sites neat and orderly, can go a long way to conserving energy and protecting the environment. A good way to remember this is Sort, Set, Shine, Standardise:

* Sort – sort and store items in your work area, eliminate clutter and manage deliveries.

* Set – everything should have its own place and be clearly marked and easy to access. In other words, be neat!

Figure 1.7 It's important to create as little waste as possible on the construction site

* Shine – clean your work area and you will be able to see potential problems far more easily.

* Standardise – by using standardised working practices you can keep organised, clean and safe.

Reducing waste is all about good working practice. By reducing wastage disposal, and recycling materials on site, you will benefit from savings on raw materials and lower transportation costs.

Planning ahead, and accurately measuring and cutting materials, means that you will be able to reduce wastage.

BASIC WORKING PLATFORMS AND ACCESS EQUIPMENT

Working at height should be eliminated or the work carried out using other methods where possible. However, there may be situations where you may need to work at height. These situations can include:

* roofing

* repair and maintenance above ground level

* working on high ceilings.

Any work at height must be carefully planned. Access equipment includes all types of ladder, scaffold and platform. You must always use a working platform that is safe. Sometimes a simple step ladder will be sufficient, but at other times you may have to use a tower scaffold.

Generally, ladders are fine for small, quick jobs of less than 30 minutes. However, for larger, longer jobs a more permanent piece of access equipment will be necessary.

Working platforms and access equipment: good practice and dangers of working at height

Table 1.10 outlines the common types of equipment used to allow you to work at heights, along with the basic safety checks necessary.

Equipment	Main features	Safety checks
Step ladder	Ideal for confined spaces. Four legs give stability	• Knee should remain below top of steps • Check hinges, cords or ropes • Position only to face work
Ladder	Ideal for basic access, short-term work. Made from aluminium, fibreglass or wood	• Check rungs, tie rods, repairs, and ropes and cords on stepladders • Ensure it is placed on firm, level ground • Angle should be no greater than 75° or 1 in 4
Mobile mini towers or scaffolds	These are usually aluminium and foldable, with lockable wheels	• Ensure the ground is even and the wheels are locked • Never move the platform while it has tools, equipment or people on it
Roof ladders and crawling boards	The roof ladder allows access while crawling boards provide a safe passage over tiles	• The ladder needs to be long enough and supported • Check boards are in good condition • Check the welds are intact • Ensure all clips function correctly
Mobile tower scaffolds	These larger versions of mini towers usually have edge protection	• Ensure the ground is even and the wheels are locked • Never move the platform while it has tools, equipment or people on it • Base width to height ratio should be no greater than 1:3
Fixed scaffolds and edge protection	Scaffolds fitted and sized to the specific job, with edge protection and guard rails	• There needs to be sufficient braces, guard rails and scaffold boards • The tubes should be level • There should be proper access using a ladder
Mobile elevated work platforms	Known as scissor lifts or cherry pickers	• Specialist training is required before use • Use guard rails and toe boards • Care needs to be taken to avoid overhead hazards such as cables

Table 1.10 Equipment for working at height and safety checks

You must be trained in the use of certain types of access equipment, like mobile scaffolds. Care needs to be taken when assembling and using access equipment. These are all examples of good practice:

* Step ladders should always rest firmly on the ground. Only use the top step if the ladder is part of a platform.

* Do not rest ladders against fragile surfaces, and always use both hands to climb. It is best if the ladder is steadied (footed) by someone at the foot of the ladder. Always maintain three points of contact – two feet and one hand.

* A roof ladder is positioned by turning it on its wheels and pushing it up the roof. It then hooks over the ridge tiles. Ensure that the access ladder to the roof is directly beside the roof ladder.

* A mobile scaffold is put together by slotting sections until the required height is reached. The working platform needs to have a suitable edge protection such as guard-rails and toe-boards. Always push from the bottom of the base and not from the top to move it, otherwise it may lean or topple over.

Figure 1.8 A tower scaffold

WORKING SAFELY WITH ELECTRICITY

It is essential whenever you work with electricity that you are competent and that you understand the common dangers. Electrical tools must be used in a safe manner on site. There are precautions that you can take to prevent possible injury, or even death.

Precautions

Whether you are using electrical tools or equipment on site, you should always remember the following:

* Use the right tool for the job.

* Use a transformer with equipment that runs on 110V.

* Keep the two voltages separate from each other. You should avoid using 230V where possible but, if you must, use a residual current device (RCD) if you have to use 230V.

* When using 110V, ensure that leads are yellow in colour.

* Check the plug is in good order.

* Confirm that the fuse is the correct rating for the equipment.

* Check the cable (including making sure that it does not present a tripping hazard).

* Find out where the mains switch is, in case you need to turn off the power in the event of an emergency.

* Never attempt to repair electrical equipment yourself.

* Disconnect from the mains power before making adjustments, such as changing a drill bit.

* Make sure that the electrical equipment has a sticker that displays a recent test date.

Visual inspection and testing is a three-stage process:

1. The user should check for potential danger signs, such as a frayed cable or cracked plug.

2. A formal visual inspection should then take place. If this is done correctly then most faults can be detected.

3. Combined inspections and **PAT** should take place at regular intervals by a competent person.

Watch out for the following causes of accidents – they would also fail a safety check:

KEY TERMS

PAT

– Portable Appliance Testing – regular testing is a health and safety requirement under the Electricity at Work Regulations (1989).

- damage to the power cable or plug
- taped joints on the cable
- wet or rusty tools and equipment
- weak external casing
- loose parts or screws
- signs of overheating
- the incorrect fuse
- lack of cord grip
- electrical wires attached to incorrect terminals
- bare wires.

When preparing to work on an electrical circuit, do not start until a permit to work has been issued by a supervisor or manager to a competent person.

Make sure the circuit is broken before you begin. A 'dead' circuit will not cause you, or anybody else, harm. These steps must be followed:

- Switch off – ensure the supply to the circuit is switched off by disconnecting the supply cables or using an isolating switch.
- Isolate – disconnect the power cables or use an isolating switch.
- Warn others – to avoid someone reconnecting the circuit, place warning signs at the isolation point.
- Lock off – this step physically prevents others from reconnecting the circuit.
- Testing – is carried out by electricians but you should be aware that it involves three parts:
  1. testing a voltmeter on a known good source (a live circuit) so you know it is working properly
  2. checking that the circuit to be worked on is dead
  3. rechecking your voltmeter on the known live source, to prove that it is still working properly.

It is important to make sure that the correct point of isolation is identified. Isolation can be next to a local isolation device, such as a plug or socket, or a circuit breaker or fuse.

The isolation should be locked off using a unique key or combination. This will prevent access to a main isolator until the work has been completed. Alternatively, the handle can be made detachable in the OFF position so that it can be physically removed once the circuit is switched off.

Dangers

You are likely to encounter a number of potential dangers when working with electricity on construction sites or in private houses. Table 1.11 outlines the most common dangers.

Danger	Identifying the danger
Faulty electrical equipment	Visually inspect for signs of damage. Equipment should be double insulated or incorporate an earth cable.
Damaged or worn cables	Check for signs of wear or damage regularly. This includes checking power tools and any wiring in the property.
Trailing cables	Cables lying on the ground, or worse, stretched too far, can present a tripping hazard. They could also be cut or damaged easily.
Cables and pipe work	Always treat services you find as though they are live. This is very important as services can be mistaken for one another. You may have been trained to use a cable and pipe locator that finds cables and metal pipes.
Buried or hidden cables	Make sure you have plans. Alternatively, use a cable and pipe locator, mark the positions, look out for signs of service connection cables or pipes and hand-dig trial holes to confirm positions.
Inadequate over-current protection	Check circuit breakers and fuses are the correct size current rating for the circuit. A qualified electrician may have to identify and label these.

Table 1.11 Common dangers when working with electricity

Each year there are around 1,000 accidents at work involving electric shocks or burns from electricity. If you are working in a construction site you are part of a group that is most at risk. Electrical accidents happen when you are working close to equipment that you think is disconnected but which is, in fact, live.

Another major danger is when electrical equipment is either misused or is faulty. Electricity can cause fires and contact with the live parts can give you an electric shock or burn you.

Different voltages

The two most common voltages that are used in the UK are 230V and 110V:

* 230V: this is the standard domestic voltage. But on construction sites it is considered to be unsafe and therefore 110V is commonly used.

* 110V: these plugs are marked with a yellow casement and they have a different shaped plug. A transformer is required to convert 230V to 110V.

Some larger homes, as well as industrial and commercial buildings, may have 415V supplies. This is the same voltage that is found on overhead electricity cables. In most houses and other buildings the voltage from these cables is reduced to 230V. This is what most electrical equipment works from. Some larger machinery actually needs 415V.

In these buildings the 415V comes into the building and then can either be used directly or it is reduced so that normal 230V appliances can be used.

Colour coded cables

Normally you will come across three differently coloured wires: Live, Neutral and Earth. These have standard colours that comply with European safety standards and to ensure that they are easily identifiable. However, in some older buildings the colours are different.

Wire type	Modern colour	Older colour
Live	Brown	Red
Neutral	Blue	Black
Earth	Yellow and Green	Yellow and Green

Table 1.12 Colour coding of cables

Working with equipment with different electrical voltages

You should always check that the electrical equipment that you are going to use is suitable for the available electrical supply. The equipment's power requirements are shown on its rating plate. The voltage from the supply needs to match the voltage that is required by the equipment.

Storing electrical equipment

Electrical equipment should be stored in dry and secure conditions. Electrical equipment should never get wet but – if it does happen – it should be dried before storage. You should always clean and adjust the equipment before connecting it to the electricity supply.

PERSONAL PROTECTIVE EQUIPMENT (PPE)

Personal protective equipment, or PPE, is a general term that is used to describe a variety of different types of clothing and equipment that aim to help protect against injuries or accidents. Some PPE you will use on a daily basis and others you may use from time to time. The type of PPE you wear depends on what you are doing and where you are. For example, the practical exercises in this book were photographed at a college, which has rules and requirements for PPE that are different to those on large construction sites. Follow your tutor's or employer's instructions at all times.

Types of PPE

PPE literally covers from head to foot. Here are the main PPE types.

Figure 1.9 A hi-vis jacket

Figure 1.10 Safety glasses and goggles

Figure 1.11 Hand protection

Figure 1.12 Head protection

Figure 1.13 Hearing protection

Protective clothing

Clothing protection such as overalls:

* provides some protection from spills, dust and irritants
* can help protect you from minor cuts and abrasions
* reduces wear to work clothing underneath.

Sometimes you may need waterproof or chemical-resistant overalls.

High visibility (hi-vis) clothing stands out against any background or in any weather conditions. It is important to wear high visibility clothing on a construction site to ensure that people can see you easily. In addition, workers should always try to wear light-coloured clothing underneath, as it is easier to see.

You need to keep your high visibility and protective clothing clean and in good condition.

Employers need to make sure that employees understand the reasons for wearing high visibility clothing and the consequences of not doing so.

Eye protection

For many jobs, it is essential to wear goggles or safety glasses to prevent small objects, such as dust, wood or metal, from getting into the eyes. As goggles tend to steam up, particularly if they are being worn with a mask, safety glasses can often be a good alternative.

Hand protection

Wearing gloves will help to prevent damage or injury to the hands or fingers. For example, general purpose gloves can prevent cuts, and rubber gloves can prevent skin irritation and inflammation, such as contact dermatitis caused by handling hazardous substances. There are many different types of gloves available, including specialist gloves for working with chemicals.

Head protection

Hard hats or safety helmets are compulsory on building sites. They can protect you from falling objects or banging your head. They need to fit well and they should be regularly inspected and checked for cracks. Worn straps mean that the helmet should be replaced, as a blow to the head can be fatal. Hard hats bear a date of manufacture and should be replaced after about 3 years.

Hearing protection

Ear defenders, such as ear protectors or plugs, aim to prevent damage to your hearing or hearing loss when you are working with loud tools or are involved in a very noisy job.

Respiratory protection

Breathing in fibre, dust or some gases could damage the lungs. Dust is a very common danger, so a dust mask, face mask or respirator may be necessary.

Make sure you have the right mask for the job. It needs to fit properly otherwise it will not give you sufficient protection.

Foot protection

Foot protection is compulsory on site, particularly if you are undertaking heavy work. Footwear should include steel toecaps (or equivalent) to protect feet against dropped objects, midsole protection (usually a steel plate) to protect against puncture or penetration from things like nails on the floor and soles with good grip to help prevent slips on wet surfaces.

Figure 1.14 Respiratory protection

Legislation covering PPE

The most important piece of legislation is the Personal Protective Equipment at Work Regulations (1992). It covers all sorts of PPE and sets out your responsibilities and those of the employer. Linked to this are the Control of Substances Hazardous to Health (2002) and the Provision and Use of Work Equipment Regulations (1992 and 1998).

Storing and maintaining PPE

All forms of PPE will be less effective if they are not properly maintained. This may mean examining the PPE and either replacing or cleaning it, or if relevant testing or repairing it. PPE needs to be stored properly so that it is not damaged, contaminated or lost. Each type of PPE should have a CE mark. This shows that it has met the necessary safety requirements.

Importance of PPE

PPE needs to be suitable for its intended use and it needs to be used in the correct way. As a worker or an employee you need to:

* make sure you are trained to use PPE

* follow your employer's instructions when using the PPE and always wear it when you are told to do so

* look after the PPE and if there is a problem with it report it.

Your employer will:

* know the risks that the PPE will either reduce or avoid

* know how the PPE should be maintained

* know its limitations.

Consequences of not using PPE

The consequences of not using PPE can be immediate or long-term. Immediate problems are more obvious, as you may injure yourself. The longer-term consequences could be ill health in the future. If your employer has provided PPE, you have a legal responsibility to wear it.

FIRE AND EMERGENCY PROCEDURES

If there is a fire or an emergency, it is vital that you raise the alarm quickly. You should leave the building or site and then head for the **assembly point.**

When there is an emergency a general alarm should sound. If you are working on a larger and more complex construction site, evacuation may begin by evacuating the area closest to the emergency. Areas will then be evacuated one-by-one to avoid congestion of the escape routes.

Three elements essential to creating a fire

Three ingredients are needed to make something combust (burn):

* oxygen　　* heat　　* fuel.

The fuel can be anything which burns, such as wood, paper or flammable liquids or gases, and oxygen is in the air around us, so all that is needed is sufficient heat to start a fire.

The fire triangle represents these three elements visually. By removing one of the three elements the fire can be prevented or extinguished.

Figure 1.15 Assembly point sign

How fire is spread

Fire can easily move from one area to another by finding more fuel. You need to consider this when you are storing or using materials on site, and be aware that untidiness can be a fire risk. For example, if there are wood shavings on the ground the fire can move across them, burning up the shavings.

Figure 1.16 The fire triangle

Heat can also transfer from one source of fuel to another. If a piece of wood is on fire and is against or close to another piece of wood, that too will catch fire and the fire will have spread.

On site, fires are classified according to the type of material that is on fire. This will determine the type of fire-fighting equipment you will need to use. The five different types of fire are shown in Table 1.13.

Class of fire	Fuel or material on fire
A	Wood, paper and textiles
B	Petrol, oil and other flammable liquids
C	LPG, propane and other flammable gases
D	Metals and metal powder
E	Electrical equipment

Table 1.13 Different classes of fire

There is also F, cooking oil, but this is less likely to be found on site, except in a kitchen.

Taking action if you discover a fire and fire evacuation procedures

During induction, you will have been shown what to do in the event of a fire and told about assembly points. These are marked by signs and somewhere on the site there will be a map showing their location.

If you discover a fire you should:

* sound the alarm

* not attempt to fight the fire unless you have had fire marshal training

* otherwise stop work, do not collect your belongings, do not run, and do not re-enter the site until the all clear has been given.

Different types of fire extinguishers

Extinguishers can be effective when tackling small localised fires. However, you must use the correct type of extinguisher. For example, putting water on an oil fire could make it explode. For this reason, you should not attempt to use a fire extinguisher unless you have had proper training.

When using an extinguisher it is important to remember the following safety points:

* Only use an extinguisher at the early stages of a fire, when it is small.

* The instructions for use appear on the extinguisher.

* If you do choose to fight the fire because it is small enough, and you are sure you know what is burning, position yourself between the fire and the exit, so that if it doesn't work you can still get out.

Type of fire risk	Fire class Symbol	White label Water	Cream label Foam	Black label Carbon dioxide	Blue label Dry powder	Yellow label Wet chemical
A – Solid (e.g. wood or paper)	A	✓	✓	✗	✓	✓
B – Liquid (e.g. petrol)	B	✗	✓	✓	✓	✗
C – Gas (e.g. propane)	C	✗	✗	✓	✓	✗
D – Metal (e.g. aluminium)	D METAL	✗	✗	✗	✓	✗
E – Electrical (i.e. any electrical equipment)	E	✗	✗	✓	✓	✗
F – Cooking oil (e.g. a chip pan)	F	✗	✗	✗	✗	✓

Table 1.14 Types of fire extinguishers

There are some differences you should be aware of when using different types of extinguisher:

- *CO_2 extinguishers* – do not touch the nozzle; simply operate by holding the handle. This is because the nozzle gets extremely cold when ejecting the CO_2, as does the canister. Fires put out with a CO_2 extinguisher may reignite, and you will need to ventilate the room after use.

- *Powder extinguishers* – these can be used on lots of kinds of fire, but can seriously reduce visibility by throwing powder into the air as well as on the fire.

SIGNS AND SAFETY NOTICES

In a well-organised working environment safety signs will warn you of potential dangers and tell you what to do to stay safe. They are used to warn you of hazards. Their purpose is to prevent accidents. Some will tell you what to do (or not to do) in particular parts of the site and some will show you where things are, such as the location of a first aid box or a fire exit.

Types of signs and safety notices

There are five basic types of safety sign, as well as signs that are a combination of two or more of these types. These are shown in Table 1.15.

Type of safety sign	What it tells you	What it looks like	Example
Prohibition sign	Tells you what you must *not* do	Usually round, in red and white	Do not use ladder
Hazard sign	Warns you about hazards	Triangular, in yellow and black	Caution Slippery floor
Mandatory sign	Tells you what you *must* do	Round, usually blue and white	Masks must be worn in this area
Safe condition or information sign	Gives important information, e.g. about where to find fire exits, assembly points or first aid kit, or about safe working practices	Green and white	First aid
Firefighting sign	Gives information about extinguishers, hydrants, hoses and fire alarm call points, etc.	Red with white lettering	Fire alarm call point
Combination sign	These have two or more of the elements of the other types of sign, e.g. hazard, prohibition and mandatory		DANGER Isolate before removing cover

Table 1.15 Different types of safety signs

TEST YOURSELF

1. Which of the following requires you to tell the HSE about any injuries or diseases?

   a. HASAWA

   b. COSHH

   c. RIDDOR

   d. PUWER

2. What is a prohibition notice?

   a. An instruction from the HSE to stop all work until a problem is dealt with

   b. A manufacturer's announcement to stop all work using faulty equipment

   c. A site contractor's decision not to use particular materials

   d. A local authority banning the use of a particular type of brick

3. Which of the following is considered a major injury?

   a. Bruising on the knee

   b. Cut

   c. Concussion

   d. Exposure to fumes

4. If there is an accident on a site who is likely to be the first to respond?

   a. First aider

   b. Police

   c. Paramedics

   d. HSE

5. Which of the following is a summary of risk assessments and is used for high risk activities?

   a. Site notice board

   b. Hazard book

   c. Monitoring statement

   d. Method statement

6. Some substances are combustible. Which of the following are examples of combustible materials?

   a. Adhesives

   b. Paints

   c. Cleaning agents

   d. All of these

7. What is dermatitis?

   a. Inflammation of the skin

   b. Inflammation of the ear

   c. Inflammation of the eye

   d. Inflammation of the nose

8. Screens, netting and guards on a site are all examples of which of the following?

   a. PPE

   b. Signs

   c. Perimeter safety

   d. Electrical equipment

9. Which of the following are also known as scissor lifts or cherry pickers?

   a. Bench saws

   b. Hand-held power tools

   c. Cement additives

   d. Mobile elevated work platforms

10. In older properties the neutral electricity wire is which colour?

    a. Black

    b. Red

    c. Blue

    d. Brown

Chapter 2

Unit CSA–L3Core07
ANALYSING TECHNICAL INFORMATION, QUANTITIES AND COMMUNICATION WITH OTHERS

LEARNING OUTCOMES

LO1: Know how to produce different types of drawings and information in the construction industry

LO2: Know how to estimate quantities and price work for contracts

LO3: Know how to ensure good working practices

INTRODUCTION

The aims of this chapter are to:

* help you to interpret information

* help you to estimate quantities

* help you to organise the building process and communicate the design work to colleagues and others.

PRODUCING DIFFERENT TYPES OF DRAWING AND INFORMATION

Accurate construction requires the creation of accurate drawings and matching supporting information. Supporting information can be found in a variety of different types of documents. These include:

* drawings and plans

* programmes of work

* procedures

* specifications

* policies

* schedules

* manufacturers' technical information

* organisational documentation

* training and development records

* risk and method statements

* Construction (Design and Management) (CDM) Regulations

* Building Regulations.

Each different type of construction plan has a definite look and purpose. Typical construction drawings focus in on floor plans or elevation.

Construction drawings are drawn to scale and need to be accurate, so that relative sizes are correct. The scale will be stated on the drawing, to avoid inaccuracies.

Working alongside these construction drawings are matched and linked specifications and schedules. It is these that outline all of the materials and tasks required to complete specific jobs.

In order to understand construction drawings you not only need to understand their purpose and what they are showing, but also a range of hatchings and symbols that act as shortcuts on the documents.

Electronic and traditional drawing methods

Construction drawings are only part of a long process in the design of buildings. In fact the construction drawings are the final stage or final version of these drawings. The design process begins with a basic concept, which is followed by outline drawings. By the end of the design stage working drawings, technical specifications and contract drawings have been completed.

The project is put out to tender. This is a process that involves companies bidding for the job based on the information that they have been given.

There are further changes just before the construction phase gets under way. The chosen construction company may have noted issues with the design, which means that the drawings may have to be amended. It is also at this stage that the construction company will begin the process of pricing up each phase of the job.

Electronic drawing methods

Many construction drawings are based on a system known as computer aided design (CAD). CAD basically produces two-dimensional electronic drawings using the similar lines, hatches and text that can be seen in traditional paper drawings.

Each different CAD drawing is created independently, so each design change has to be followed up on other CAD drawings.

Increasingly, however, a new electronic system is being used. It is known as building information modelling (BIM). This creates drawings in 3D. The buildings are virtually modelled from real construction elements, such as walls, windows and roofs. The big advantages are:

* it allows architects to design buildings in a similar way to the way in which the building will actually be built

* a central virtual building model stores all the data, so any changes to this are applied to individual drawings

* better coordinated designs can be created meaning that construction should be more straightforward.

Systems such as BIM provide 3D models, which can be viewed from any angle or perspective. It also includes:

* scheduling information

* labour required

* estimated costs

* a detailed breakdown of the construction phases.

Figure 2.1 BIM generated model

Traditional drawing methods

The development of the computer, laptops and hand-held tablets such as iPads is gradually making manual drafting of construction drawings obsolete. The majority of drawings are now created using CAD or BIM software.

Traditionally, drawings were limited to the available paper size and what would be convenient to transport.

As each of the traditional construction drawings were hand drawn, there was always a danger that the information on one drawing would not match the information on another. The only way to check that both were accurate was to cross-reference every detail.

One advantage is that paper plans are easier to carry around site, as computers can be broken or stolen. However, damaging or losing a paper plan can cause delays while it is replaced.

Types of supporting information

Drawings and plans

Drawings are an important part of construction work. You will need to understand how they provide you with the information you need to carry out the work. The drawings show what the building will look like and how it will be constructed. This means that there are several different drawings of the building from different viewpoints. In practice most of the drawings are shown on the same sheet.

Block plans

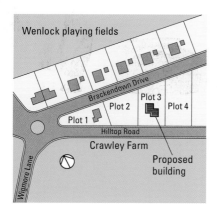

Figure 2.2 Block plan

Block plans show the construction site and the surrounding area. Normally block plans are at a ratio of 1:1250 and 1:2500. This means that 1 millimetre on a block plan is equal to 1,250 mm (12.5 m) or 2,500 mm (25 m) or on the ground.

Site plan

Often location drawings are also known as block plans or site plans. The site plan drawing shows what is planned for the site. It is often an important drawing because it has been created in order to get approval for the project from planning committees or funding sources. In most cases the site plan is actually an architectural plan, showing the basic arrangement of buildings and any landscaping.

The site plan will usually show:

* directional orientation (i.e. the north point)

* location and size of the building or buildings

* existing structures

* clear measurements.

General location

Location drawings show the site or building in relation to its surroundings. It will therefore show details such as boundaries, other buildings and roads. It will also contain other vital information, including:

* access

* drainage

* sewers

* the north point.

As with all plan drawings, the scale will be shown and the drawing will be given a title. It will be given a job or project number to help identify it easily, as well as an address, the date of the drawing and the name of the client. A version number will also be on the drawing with an amendment date if there have been any changes. You'll need to make sure you have the latest drawing.

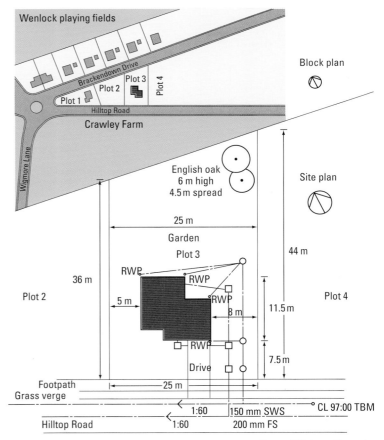

Figure 2.3 Location plan

Normally location drawings are either 1:200 or 1:500 (that is, 1 mm of the drawing represents 200 mm (2 m) or 500 mm (5 m) on the ground).

Assembly

These are detailed drawings that illustrate the different elements and components of the construction. They tend to be 1:5, 1:10 or 1:20 (1 mm of the drawing represents 5, 10 or 20 mm on the ground). This larger scale allows more detail to be shown, to ensure accurate construction.

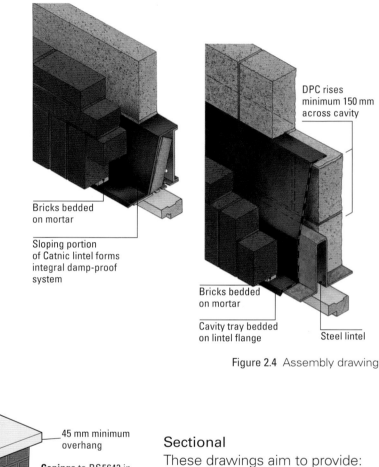

DPC rises minimum 150 mm across cavity

Bricks bedded on mortar

Sloping portion of Catnic lintel forms integral damp-proof system

Bricks bedded on mortar

Cavity tray bedded on lintel flange

Steel lintel

Figure 2.4 Assembly drawing

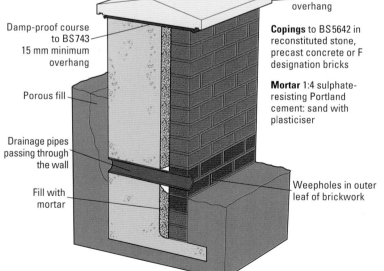

45 mm minimum overhang

Damp-proof course to BS 743 15 mm minimum overhang

Porous fill

Drainage pipes passing through the wall

Fill with mortar

Copings to BS 5642 in reconstituted stone, precast concrete or F designation bricks

Mortar 1:4 sulphate-resisting Portland cement: sand with plasticiser

Weepholes in outer leaf of brickwork

Sectional

These drawings aim to provide:

* vertical and horizontal measurements and details

* constructional details.

They can be used to show the height of ground levels, damp-proof courses, foundations and other aspects of the construction.

Figure 2.5 Section drawing of an earth retaining wall

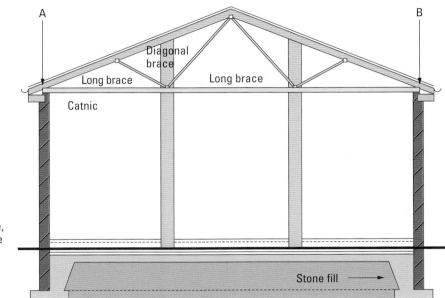

Figure 2.6 Section drawing of a garage

Details

These drawings show how a component needs to be manufactured. They can be shown in various scales, but mainly 1:10, 1:5 and 1:1 (the same size as the actual component if it is small).

Programmes of work

Programmes of work show the actual sequence of any work activities on a construction project. Part of the work programme plan is to show target times. They are usually shown in the form of a Gantt chart (a special type of bar chart), as can be seen in Fig 2.8.

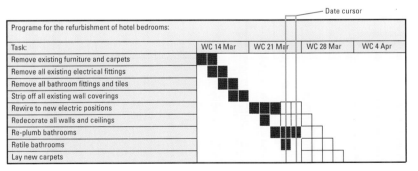

Figure 2.8 Single line contract plan Gantt chart

In this figure:

● on the left-hand side all of the tasks are listed – note this is in logical order

● on the right the blocks show the target start and end date for each of the individual tasks

● the timescale can be days, weeks or months.

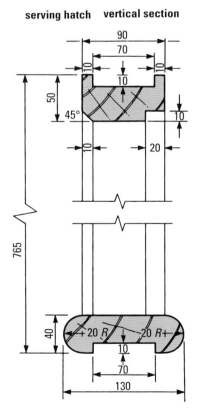

Figure 2.7 Detail drawing

Far more complex forms of work programmes can also be created. Fig 2.9 shows the planning for the construction of a house.

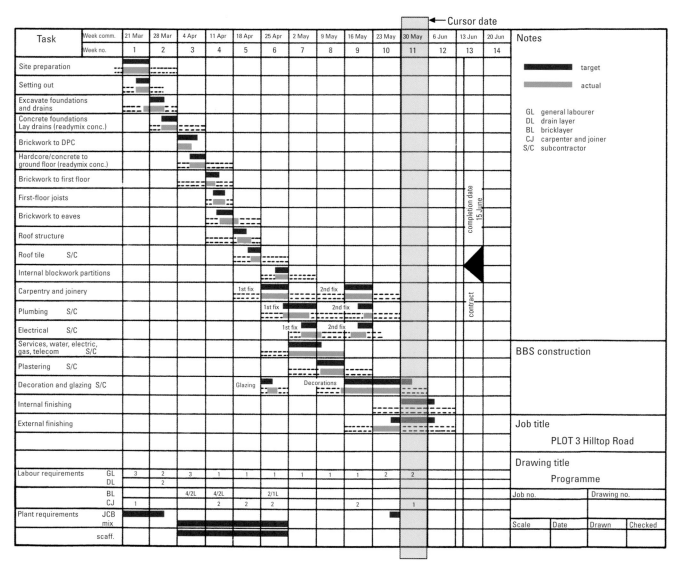

Figure 2.9 Gantt chart for the construction of a house

This more complex example shows the following:

* There are two lines – they show the target dates and actual dates. The actual dates are shaded, showing when the work actually began and how long it took.

* If this bar chart is kept up to date an accurate picture of progress and estimated completion time can be seen.

Procedures

When you work for a construction company they will have a series of procedures they will expect you to follow. A good example is the emergency procedure. This will explain precisely what is required in the case of an emergency on site and who will have responsibility to carry out particular duties. Procedures are there to show you the right way of doing something.

Another good example of a procedure is the procurement or buying procedure. This will outline:

* who is authorised to buy what, and how much individuals are allowed to spend

* any forms or documents that have to be completed when buying.

Specifications

In addition to drawings it is usually necessary to have documents known as specifications. These provide much more information, as can be seen in Fig 2.10.

The specifications give you a precise description. They will include:

* the address and description of the site

* on-site services (e.g. water and electricity)

* materials description, outlining the size, finish, quality and tolerances

* specific requirements, such as the individual that will authorise or approve work carried out

* any restrictions on site, such as working hours.

Policies

Policies are sets of principles or a programme of actions. The following are two good examples:

* environmental policy – how the business goes about protecting the environment

* safety policy – how the business deals with health and safety matters and who is responsible for monitoring and maintaining it.

You will normally find both policies and procedures in site rules. These are usually explained to each new employee when they first join the company. Sometimes there may be additional site rules, depending on the job and the location of the work.

Schedules

Schedules are cross-referenced to drawings that have been prepared by an architect. They will show specific design information. Usually they are prepared for jobs that will crop up regularly on site, such as:

* working on windows, doors, floors, walls or ceilings

* working on drainage, lintels or sanitary ware.

BBS DESIGN

Specification of the works to be carried out and the materials to be used in the erection and completion of a new house and garage on plot 3, Hilltop Road, Brackendowns, Bedfordshire, for Mr W. Whiteman, to the satisfaction of the architect.

1.00   General conditions

1.01
1.02

| 10.00 | Woodwork |

1.03
1.04

10.01   Timber for carcassing work to be machine strength graded class C16

2.00
2.01
2.02

10.02   Timber for joinery shall be a species approved by the architect and specified as J2

2.03
2.04
2.05

2.06
2.07
2.08
2.09

10.03   Moisture content of all timber at time of fixing to be appropriate to the situation and conditions in which it is used. To this effect all timber and components will be protected from the weather prior to their use.

10.04

10.18   Construct the first floor using 50 mm × 195 mm sawn softwood joists at 400 mm centres supported on mild steel hangers.

10.05

Provide 75 mm × 195 mm trimmer and trimming around stairwell, securely tusk-tenoned together.

Provide and fix to joists 38 mm × 38 mm sawn softwood herring-bone strutting at 1.8 m maximum intervals.

10.06

Provide and fix galvanized restraint straps at 2 m maximum intervals to act as positive ties between the joists and walls.

10.19   Provide and secret fix around the trimmed stairwell opening a 25 mm Brazilian mahogany apron lining, tongued to a matching 25 mm × 100 mm nosing.

10.20   Provide and lay to the whole of the first floor 19 mm × 100 mm prepared softwood tongued and grooved floor boarding, each board well cramped up and surface nailed with two 50 mm flooring brads to each joist. The nail heads to be well punched down.

Figure 2.10 Extracts from a typical specification

A schedule can be seen in Fig 2.11.

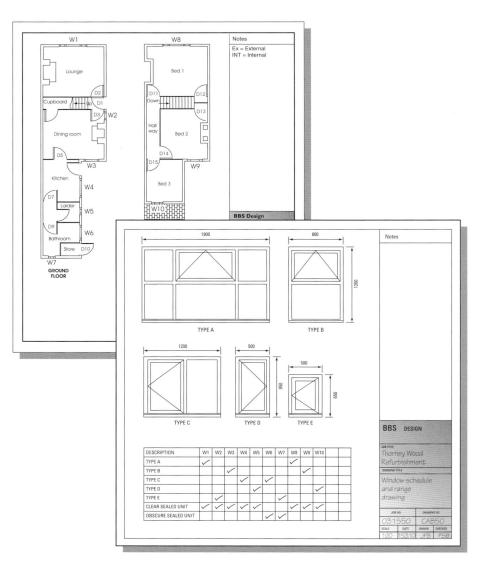

Figure 2.11 Typical windows schedule, range drawing and floor plans

The schedule is very useful for a number of reasons:

● working out the quantities of materials needed

● ordering materials and components and then checking them against deliveries

● locating where specific materials will be used.

Manufacturers' technical information

Almost everything that is bought to be used on site will come with a variety of information. The basic technical information provided will show what the equipment or material is intended to be used for, how it should be stored and any particular requirements it may have, such as handling or maintenance.

Technical information from the manufacturer can come from a variety of different sources:

* printed or downloadable data sheets

* printed or downloadable user instructions

* manufacturers' catalogues or brochures

* manufacturers' websites.

Organisational documentation

The potential list of organisational documentation and paperwork is massive. Examples are outlined in the following table.

Document	Purpose
Timesheet	Record of hours that you have worked and the jobs that you have carried out. They are used to help work out your wages and the total cost of the job.
Day worksheet	These detail work that has been carried out without providing an estimate beforehand. They usually include repairs or extra work and alterations.
Variation order	These are provided by the architect and given to the builder, showing any alterations, additions or omissions to the original job.
Confirmation notice	Provided by the architect to confirm any verbal instructions.
Daily report or site diary	Include things that might affect the project like detailed weather conditions, late deliveries or site visitors.
Orders and requisitions	These are order forms, requesting the delivery of materials.
Delivery notes	These are provided by the supplier of materials as a list of all materials being delivered. These need to be checked against materials actually delivered.
Delivery record	These are lists of all materials that have been delivered on site.
Memorandum	These are used for internal communications and are usually brief.
Letters	These are used for external communications, usually to customers or suppliers.
Fax	Even though email is commonly used, the industry still likes faxes, because they provide an exact copy of an original document.

Table 2.1

Training and development records

Training and development is an important part of any job, as it ensures that employees have all the skills and knowledge that they need to do their work. Most medium to large employers will have training policies that set out how they intend to do this.

To make sure that they are on track and to keep records they will have a range of different documents. These will record all the training that an employee has undertaken.

Training can take place in a number of different ways:

* induction
* toolbox talks
* in-house training
* specialist training
* training or education leading to formal qualifications.

Details required for floor plans

The floor plans shows the arrangement of the building, rather like a map. It is a cut through of the building, which shows openings, walls and other features usually at around 1 m above floor level.

The floor plan also includes elements of the building that can be seen below the 1 m level, such as the floor or part of the stairs. The drawing will show elements above the 1 m level as dotted lines. The floor plan is a vertical orthographic projection onto a horizontal plane. In effect the horizontal plane cuts through the building.

The floor plan will detail the following:

* Vertical and horizontal sections – these show the building cut along an axis to show the interior structure.

* Datum levels – these are taken from a nearby and convenient datum point. They show the building's levels in relation to the datum point.

* Wall constructions – this is revealed through the section or cross sections shown in the diagram. It details the wall construction methods and materials.

* Material codes – these will contain notes and links to specific materials and may also note particular parts of the Building Regulations that these construction materials comply with.

* Depth and height dimensions – these are drawn between the walls to show the room sizes and wall lengths. They are noted as width × depth.

* Schedules – these note repeated design information, such as types of door, windows and other features.

* Specifications – these outline the type, size and quality of materials, methods of fixing and quality of work and finish expected.

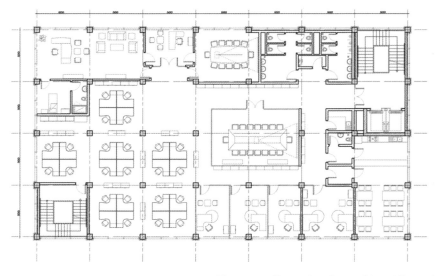

Figure 2.12 Example of a traditional floor plan

Details required for elevations

The details required for elevations in construction drawings are the same as those required for floor plans. An elevation is the view of the building as seen from one side. It can be used to show what the exterior of the building will look like. The elevation is labelled in relation to the compass direction. The elevation is a horizontal orthographic projection of the building onto a vertical plane. Usually the vertical plane is parallel to one side of the building (orthographic drawing).

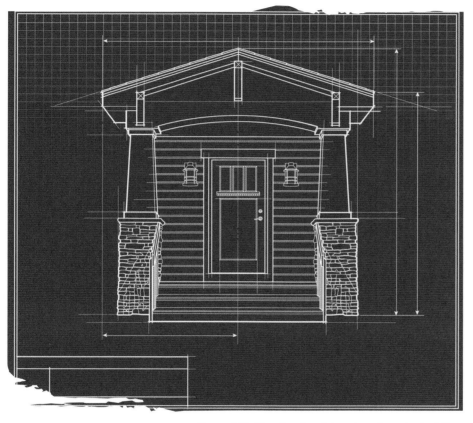

Figure 2.13 The details of the elevation of a building

Linking schedules to drawings

The schedule of work and the drawings create a single set of information. These documents need to be clear and comprehensive.

Before construction gets under way the specification schedule is the most important set of documents. It is used by the construction company to price up the job, work out how to tackle it, and then put in a bid for the work.

The construction company can look at each task in detail and see what materials are needed. This, along with all the construction information documents, will help them to make an estimate as to how long the task will take to complete and to what standard it should be completed.

During the construction period the most important documents are the drawings. Each piece of work is linked to those drawings and a schedule of work is set up. This might incorporate a Gantt chart or critical path analysis, showing expected dates and duration of on-site and off-site activities. This might need a good deal of cross-referencing. Obviously you cannot fit windows until the relevant cavity wall has been built and the opening formed.

It is important that the drawings and the specification schedules are closely linked. Reference numbers and headings that are on the drawings need to appear with exactly the same numbers and words on the schedule. This will avoid any confusion. It should be possible to look at the drawings, find a reference number or heading and then look through the schedule to find the details of that particular task. It also allows the drawings to be slightly clearer, as they won't need to have detailed information on them that can be found in the schedule.

Reasons for different projections in construction drawings

Designers will use a range of drawings in order to get across their requirements. Each is a 2D image. They show what the building will look like, along with the components or layout.

Orthographic projections
Orthographic projections are used to show the different elevations or views of an object. Each of the views is at right angles to the face.

PRACTICAL TIP

An orthographic projection is a way of illustrating a 3D object like a building in 2D.

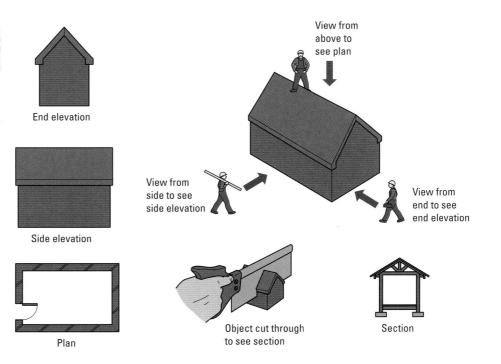

End elevation

Side elevation

Plan

View from above to see plan

View from side to see side elevation

View from end to see end elevation

Object cut through to see section

Section

Figure 2.14 Plans, elevations and sections

Orthographic projection can be seen either as a first angle European projection or a third angle American projection. The following table shows the difference between these two views and there are examples in Figs 2.16 and 2.17, which relate to the shape shown in Fig 2.15.

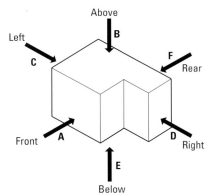

Figure 2.15 Isometric diagram showing the various views that can be portrayed in orthographic projection

Projection	Description
First angle	Everything is drawn in relation to the front view. The view from above is drawn below and the view from below is drawn from above. The view from the left is to the right and the right to the left. So all views, in effect, are reversed.
Third angle	This is often referred to as being an American projection. Everything again is in relation to the front elevation. The views from above and below are drawn in their correct position. Anything on the left is drawn to the left and the right to the right.

Table 2.2

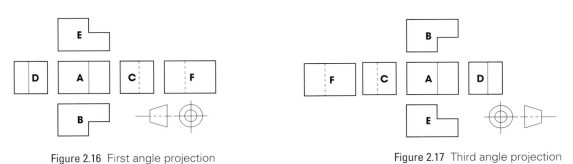

Figure 2.16 First angle projection

Figure 2.17 Third angle projection

Pictorial projections

Pictorial projections show objects in a 3D form. There are different ways of showing the view by varying the angles of the base line and the scale of any side projections. The most common is isometric. Vertical lines are drawn vertically, and horizontal lines are drawn at an angle of 30° to the horizontal. All of the other measurements are drawn to the same scale. This type of pictorial projection can be seen in Fig 2.18.

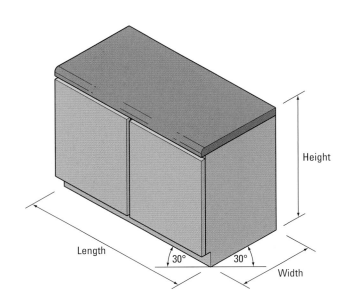

Figure 2.18 Isometric projection

There are four other different types of pictorial projection. These are not used as commonly as isometric projections.

Pictorial projection	Description
Planometric	Vertical lines are drawn vertically and horizontal lines on the front elevation of the object are drawn at 30°. The horizontal lines on the side elevation are drawn at 60° to horizontal.
Axonometric	The horizontal lines on all elevations are drawn at 45° to the horizontal. Otherwise the look is very similar to planometric.
Oblique	All of the vertical lines are drawn vertically. The horizontal lines on the front elevation are drawn horizontally but all the other horizontal lines are drawn at 45° to the horizontal.
Perspective	Horizontal lines are drawn so that they disappear into an imaginary horizon, known as a vanishing point. A one-point perspective drawing has all the sides disappearing to one vanishing point. An angular perspective, or two point perspective, has the elevations disappearing to two vanishing points.

Table 2.3

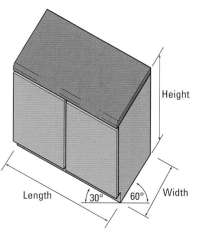

Figure 2.19 Planometric projection

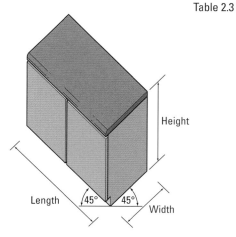

Figure 2.20 Axonometric projection

Figure 2.21 Oblique projection

VP = viewpoint

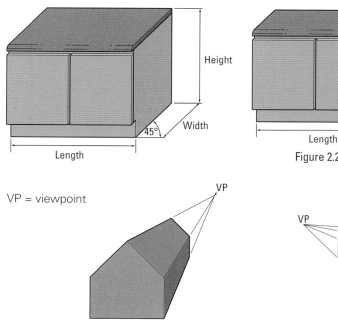

Figure 2.22 Parallel (one point) perspective projection

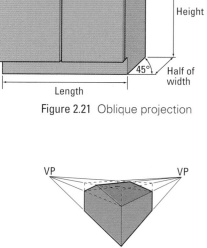

Figure 2.23 Angular (two point) perspective projection

Hatchings and symbols

Different materials and components are shown using symbols and hatchings. Abbreviations are also used. This makes the working drawings far less cluttered and easier to read.

Examples of symbols and abbreviations can be seen in Figs 2.24 and 2.25.

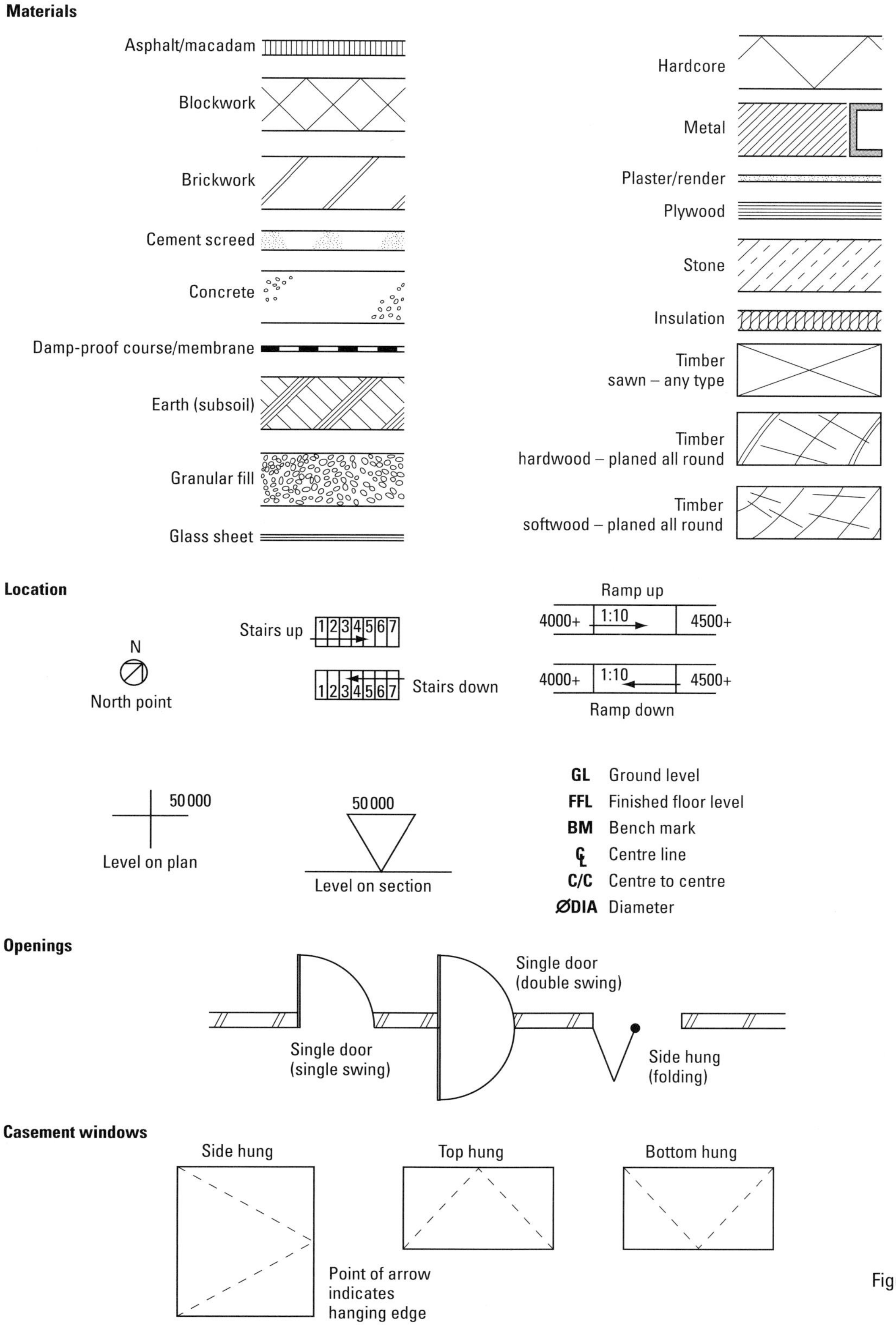

Materials

Asphalt/macadam

Blockwork

Brickwork

Cement screed

Concrete

Damp-proof course/membrane

Earth (subsoil)

Granular fill

Glass sheet

Hardcore

Metal

Plaster/render

Plywood

Stone

Insulation

Timber
sawn – any type

Timber
hardwood – planed all round

Timber
softwood – planed all round

Location

N

North point

Stairs up 1234567

1234567 Stairs down

Ramp up

4000+ 1:10 4500+

4000+ 1:10 4500+

Ramp down

50 000

Level on plan

50 000

Level on section

GL	Ground level
FFL	Finished floor level
BM	Bench mark
₵	Centre line
C/C	Centre to centre
ØDIA	Diameter

Openings

Single door
(single swing)

Single door
(double swing)

Side hung
(folding)

Casement windows

Side hung

Top hung

Bottom hung

Point of arrow
indicates
hanging edge

Figure 2.24 Symbols used on drawings

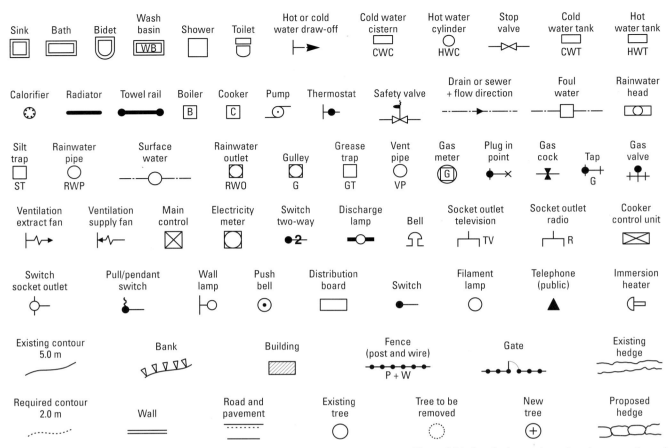

Figure 2.24 Symbols used on drawings *continued*

Aggregate	agg	BS tee	BST	Foundation	fdn	Polyvinyl acetate	PVA
Air brick	AB	Building	bldg	Fresh air inlet	FAI	Polyvinylchloride	PVC
Aluminium	al	Cast iron	CI	Glazed pipe	GP	Rainwater head	RWH
Asbestos	abs	Cement	ct	Granolithic	grano	Rainwater pipe	RWP
Asbestos cement	absct	Cleaning eye	CE	Hardcore	hc	Reinforced concrete	RC
Asphalt	asph	Column	col	Hardboard	hdbd	Rodding eye	RE
Bitumen	bit	Concrete	conc	Hardwood	hwd	Foul water sewer	FWS
Boarding	bdg	Copper	Copp cu	Inspection chamber	IC	Surface water sewer	SWS
Brickwork	bwk	Cupboard	cpd	Insulation	insul	Softwood	swd
BS* Beam	BSB	Damp-proof course	DPC	Invert	inv	Tongued and grooved	T&G
BS Universal beam	BSUB	Damp-proof membrane	DPM	Joist	jst	Unglazed pipe	UGP
BS Channel	BSC	Discharge pipe	DP	Mild steel	MS	Vent pipe	VP
BS equal angle	BSEA	Drawing	dwg	Pitch fibre	PF	Wrought iron	WI
BS unequal angle	BSUA	Expanding metal lathing	EML	Plasterboard	pbd		

Figure 2.25 Abbreviations commonly used on drawings

ESTIMATING QUANTITIES AND PRICING WORK FOR CONTRACTS

Working out the quantity and cost of resources that are needed to do a particular job can be difficult. In most cases you or the company you work for will be asked to provide a price for the work. It is generally accepted that there are three ways of doing this:

* Estimate – which is an approximate price, though estimation is a skill based on many factors.

* Quotation – which is a fixed price.

* Tender – which is a competitive quotation against other companies for a prescribed amount of work to a certain standard.

As we will see a little later in this section, these three ways of costing are very different and each of them has its own issues.

Resource requirements

As you become more experienced you will be able to estimate the amount of materials that will be needed on particular construction projects though this depends on the size and complexity of the job. This is also true of working out the best place to buy materials and how much the labour costs will be to get the job finished.

In order to work out how much a job will cost, you will need to know some basic information:

* What type of contract is agreed?

* What materials will be used?

* What are the costs of the materials?

Much of this information can be gained from the drawings, specification and other construction information for the proposed building.

To help work out the price of a job, many businesses use the *UK Building Blackbook,* which provides a construction cost guide. It breaks down all types of work and shows an average cost for each of them.

Computerised estimating packages are available, which will give a comprehensive detailed estimate that looks very professional. This will also help to estimate quantities and timescales.

Measurement
The standard unit for measurement is based on the metre (m). There are 100 centimetres (cm) and 1,000 millimetres (mm) in a metre. It is important to remember that drawings and plans have different scales, so these need to be converted to work out quantities of materials.

The most basic thing to work out is length, from which you can calculate perimeter, area and then volume, capacity, mass and weight, as can be seen in the following table.

Measurement	Explanation
Length	This is the distance from one end to the other. For most jobs metres will be sufficient, although for smaller work such as brick length or lengths of screws, millimetres are used.
Perimeter	This is the distance around a shape, such as the size of a room or a garden. It will help you estimate the length of a wall, for example. You just need to measure each side and then add them together.
Area	You can work out the area of a room, for example, by measuring its length and its width. Then you multiply the width by the length to give the number of square metres (m^2).
Volume and capacity	Volume shows how much space is taken up by an object, such as a room. Again this is simply worked out by multiplying the width of the room by its length and then by its height. This gives you the number of cubic metres (m^3). Capacity works in exactly the same way but instead of showing the figure as cubic metres you show it as litres. This is ideal if you are trying to work out the capacity of the water tank or a garden pond.
Mass or weight	Mass is measured usually in kilograms or in grams. Mass is the actual weight of a particular object, such as a brick.

Table 2.4

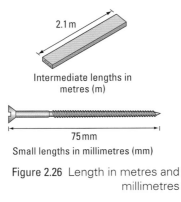

Figure 2.26 Length in metres and millimetres

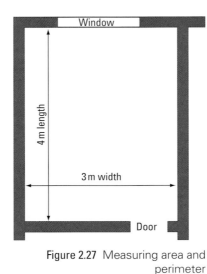

Figure 2.27 Measuring area and perimeter

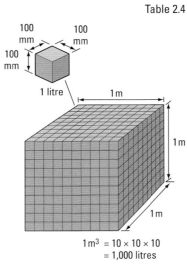

Figure 2.28 Relationship between volume and capacity

Formulae

These can appear to be complicated, but using formulae is essential for working out quantities of materials. Each formula is related to different shapes. In construction you will often have to work out quantities of materials needed for odd shaped areas.

Area

To work out the area of a triangular shape, you use the following formula:

$$Area (A) = Base (B) \times Height (H) \div 2$$

So if a triangle has a base of 4.5 and a height of 3.5 the calculation is:

$$4.5 \times 3.5 \div 2$$

Or $4.5 \times 3.5 = 15.75 \div 2 = 7.875\,m^2$

Height

If you want to work out the height of a triangle you switch the formulae around. To give us height = 2 × Area ÷ Base

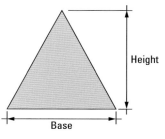

Figure 2.29 Triangle

Perimeter

To work out the perimeter of a rectangle use the formula:

$$Perimeter = 2 \times (Length + Width)$$

It is important to remember this because you need to count the length and the width twice to ensure you have calculated the total distance around the object.

Circles

To work out the circumference or perimeter of a circle you use the formula:

$$Circumference = \pi\ (pi) \times diameter$$

π (pi) is always the same for all circles and is 3.142.

Diameter is the length of the widest part.

If you know the circumference and need to work out the diameter of the circle the formula is:

$$Diameter = circumference \div \pi\ (pi)$$

For example if a circle has a circumference of 15.39 m then to work out the diameter:

$$15.39 \div 3.142 = 4.89\,m$$

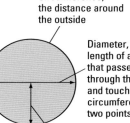

Circumference, the distance around the outside

Diameter, the length of a line that passes through the centre and touches the circumference at two points

Radius, half the diameter

Figure 2.30 Parts of a circle

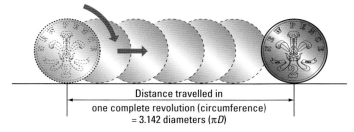

Distance travelled in one complete revolution (circumference) = 3.142 diameters (πD)

Figure 2.31 Relationship between circumference and diameter

Complex areas

Land, for example, is rarely square or rectangular. It is made up of odd shapes. Never be overwhelmed by complex areas, as all you need to do is to break them down into regular shapes.

By accurately measuring the perimeter you can then break down the shape into a series of triangles or rectangles. All you need to do then is to work out the area of each of the shapes within the overall shape and then add them together.

Shape		Area equals	Perimeter equals
Square		AA (or A multiplied by A)	4A (or A multiplied by 4)
Rectangle		LB (or L multiplied by B)	2(L + B) (or L plus B multiplied by 2)
Trapezium		$\dfrac{(A + B)H}{2}$ (or A plus B multiplied by H and then divided by 2)	A + B + C + D
Triangle		$\dfrac{BH}{2}$ (or B multiplied by H and then divided by 2)	A + B + C
Circle		πR^2 (or R multiplied by itself and then multiplied by pi (3.142))	πD or $2\pi R$

Figure 2.32 Table of shapes and formulae

Volume

Sometimes it is necessary to work out the volume of an object, such as a cylinder or the amount of concrete needed. All that needs to be done is to work out the base area and then multiply that by the height.

For a concrete area, if a 1.2 m square needs 3 m of height then the calculation is:

$$1.2 \times 1.2 \times 3 = 4.32\,\text{m}^3$$

To work out the volume of a cylinder you need to know the base area × the height. The formula is:

$$\pi r^2 \times H$$

So if a cylinder has a radius (r) of 0.8 and a height of 3.5 m then the calculation is:

$$3.142 \times 0.8 \times 0.8 \times 3.5 = 7.038\,\text{m}^3$$

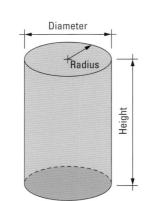

Figure 2.33 Cylinder

Pythagoras

Pythagoras' theorem is used to work out the length of the sides of right angled triangles. The theory states that:

In all right angled triangles the square of the longest side is equal to the sum of the squares of the other two sides (that is, the length of a side multiplied by itself).

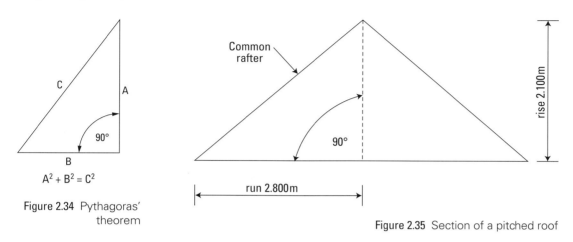

$$A^2 + B^2 = C^2$$

Figure 2.34 Pythagoras' theorem

Figure 2.35 Section of a pitched roof

See Chapter 7 for more about geometry, trigonometry and Pythagoras.

Measuring materials

Using simple measurements and formulae can help you work out the amount of materials you will need. This is all summarised in the following table.

Material	Measurement
Timber	Can be sold by the cubic metre. To work out the length of material divide the cross section area of one section by the total cross section area of the material.
Flooring	To work out the amount of flooring for a particular area multiply the width of the floor by the length of the floor.
Stud walling	Measure the distance that the stud partition will cover then divide that distance by a specified spacing. This will give you the number of spaces between each stud.
Rafters and floor joists	Measure the distance between the adjacent walls then take into account that the first and last joist or rafter will be 50 mm away from the wall. Measure the total distance and then divide it by the specified spacing.
Fascias, barges and soffits	Measure the length and then add a little extra to take into account any necessary cutting and jointing.
Skirting	You need to work out the perimeter of the room and then subtract any doorways or other openings. This technique can be used to work out the necessary length of dado, picture rails and coving.
Bricks and mortar	Half-brick walls use 60 bricks per metre squared and one-brick walls use double that amount. You should add 5 per cent to take into account any cutting or damage. For mortar assume that you will need 1 kg for each brick.

Table 2.5

How to cost materials

Once you have found out the quantity of materials necessary you will need to find out the price of those materials. It is then simply a case of multiplying those prices by the amount of materials actually needed to find out approximately how much they will cost in total.

Materials and purchasing systems

Many builders and companies will have preferred suppliers of materials. Many of them will already have negotiated discounts based on their likely spending with that supplier over the course of a year. The supplier will be geared up to supply them at an agreed price.

In other cases builders may shop around to find the best price for the materials that match the specification. It is not always the case that the lowest price is necessarily the best. All materials need to be of a sufficient quality. The other key consideration is whether the materials are immediately available for delivery.

It is vital that suppliers are reliable and that they have sufficient materials in stock. Delays in deliveries can cause major setbacks on site. It is not always possible to warn suppliers that materials will be needed, but a well-run site should be able to anticipate the materials that are needed and put in the orders within good time.

Large quantities may be delivered direct from the manufacturer straight to site. This is preferable when dealing with items where consistency, for example of colour, is required.

Labour rates and costs

The cost of labour for particular jobs is based on the hourly charge-out rate for that individual or group of individuals multiplied by the time it would take to complete the job.

Labour rates can depend on the:

* expertise of the construction worker

* size of the business they work for

* part of the country in which the work is being carried out

* complexity of the work.

According to the International Construction Costs Survey 2012, the following were average costs per hour:

* Group 1 tradespeople – plumbers, electricians etc.: £30

* Group 2 tradespeople – carpenters, bricklayers etc.: £30

* Group 3 tradespeople – tillers, carpet layers and plasterers: £30

* general labourers: £18

* site supervisors: £46.

REED TIP

A great career path can start with an apprenticeship. 80 per cent of the staff at South Tyneside Homes started off as apprentices. Some have worked their way up to job roles such as team leaders, managers and heads of departments.

Quotes, estimated prices and tenders

As we have already seen, estimates, quotes and tenders are very different. We need to look at these in slightly more detail, as can be seen in the following table.

Type of costing	Explanation
Estimate	This needs to be a realistic and accurate calculation based on all the information available as to how much a job will cost. An estimate is not binding and the client needs to understand that the final cost might be more.
Quote	This is a fixed price based on a fixed specification. The final price may be different if the fixed specification changes; for example if the customer asks for additional work then the price will be higher.
Tender	This is a competitive process. The customer advertises the fact that they want a job done and invites tenders. The customer will specify the specifications and schedules and may even provide the drawings. The companies tendering then prepare their own documents and submit their price based on the information the customer has given them. All tenders are submitted to the customer by a particular date and are sealed. The customer then opens all tenders on a given date and awards the contract to the company of their choice. This process is particularly common among public sector customers such as local authorities.

Table 2.6

Inaccurate estimates

Larger companies will have an estimating team. Smaller businesses will have someone who has the job of being an estimator. Whenever they are pricing a job, whether it is a quote, an estimate or a tender, they will have to work out the costs of all materials, labour and other costs. They will also have to include a **mark-up.**

It is vital that all estimating is accurate. Everything needs to be measured and checked. All calculations need to be double-checked.

It can be disastrous if these figures are wrong because:

* if the figure is too high then the client is likely to reject the estimate and look elsewhere as some competitors could be cheaper

* if the figure is too low then the job may not provide the business with sufficient profit and it will be a struggle to make any money out of the job.

KEY TERMS

Mark-up

– a builder or building business, just like any other business, needs to make a profit. Mark-up is the difference between the total cost of the job and the price that the customer is asked to pay for the work.

DID YOU KNOW?

Many businesses fail as a result of not working out their costs properly. They may have plenty of work but they are making very little money.

CASE STUDY

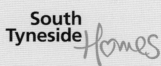

South Tyneside Council's

Bringing all your skills together to do a good job

Marcus Chadwick, a bricklayer at Laing O'Rourke, talks about maths and English skills.

'Obviously you need your maths, especially being a bricklayer. From the dimensions on your drawings, you have to be able to work out how many materials you'll need – how many bricks, how many blocks, how much sand and cement you need to order. Eventually it ends up being rote learning, like the way you learn your times-tables. With a bit of practice, you'll be able to work out straight away, "Right, I need x number of blocks, I need 1000 bricks there, I need a ton of sand, therefore I need seven bags of cement." Though there are still times when I get the calculator out!

If you get it wrong and miscalculate it can delay the progression of the building, or your section. I'm the foreman and if I set out a wall in the wrong position then there's only one person to blame. So you check, then double-check – it's like the old saying, "Measure twice, cut once".

Number skills really are important; you can't just say "Well, I'm a bricklayer and I'm just going to work with my hands". But that all comes with time; I wasn't that good at maths when I left school, though it was probably a case of just being lazy. When you come into an environment where you need to start using it to earn the money, then you'll start to get it straight away.

Your communication skills are definitely important too. You need to know how to speak to people. I always find that your lads appreciate you more if you ask them to do something as opposed to tell them. That was your old 1970s mentality where you used to have your screaming foreman – "Get this done, get that done!" – it doesn't work like that anymore. You have to know how to speak to people, to communicate.

All the lads that work for me, they're my extended family. My boss knows that too and that's why I've been with him for seven years now – he takes me everywhere because he knows I've got a good relationship with all my workforce.

You'll also have the odd occasion where the client will come around to visit the site and you've got to be able to put yourself across, using good diction. That also goes for when you're ordering materials – because you deal with different regions across the country, you've got to be clearly spoken so they understand you, so things don't get messed up in translation.'

Purchasing or hiring plant and equipment

Normally, if a piece of plant or equipment is going to be used on a regular basis then it is purchased by the company. By maximising the use of any plant or equipment, the business will save on the costs of repeatedly hiring and the transport of the item to and from the site.

It also does not make sense to leave plant and equipment on a site if it is no longer being used. It needs to be moved to a new site where it can be used.

Many smaller construction companies have no alternative other than to hire. This is because they cannot afford to have an enormous amount of money tied up in the plant or equipment, whether this comes from earned profits or from a loan or finance agreement. Loans and finance agreements have to be paid back over a period of time.

The decision as to whether to purchase or to hire is influenced by a number of factors:

* The working lives of the plant or equipment – how long will it last? This will usually depend on how much it is used and how well maintained it is.

* The use of the plant or equipment – is the company going to get good use out of it if they buy it? If they are hiring it then it should only be hired for the time it is actually needed. There is no point in having the plant or equipment on site and paying for its hire if it is not being used.

* Loss of value – just like buying a brand new car, the value of new plant or equipment takes an enormous drop the moment you take delivery of it. Even if it is hardly used it is considered second-hand and is not worth anything like its price when it was new. The biggest falls in value are in the first few years that the company owns it. It then reaches a value that it will sit at for some years until it is considered junk or scrap.

* Obsolescence – what might seem today to be the most advanced and technologically superior piece of plant or equipment may not be so tomorrow. Newer versions will come onto the market and may be more efficient or cost-effective. It is probable that the plant or equipment will be obsolete, or outdated, before it ends its useful working life.

* Cost of replacement – investing in plant or equipment today means that at some point in the future they will have to be replaced. The business will have to take account of this and arrange to have the necessary funds available for replacement in the future.

* Maintenance costs – if the construction business owns the plant or equipment they will have to pay for any routine maintenance, repairs and of course operators. Hired equipment, such as diggers or cranes, is the responsibility of the hiring company. They pay for all the maintenance and although they charge for the operator, the operator is on their wage bill.

* Insurance and licences – owning plant or equipment often means additional insurance payments and the company may also have to obtain licences that allow them to use that type of equipment in a particular area. Hired plant and equipment is already insured and should have the relevant licences.

* Financial costs – if the decision is to buy rather than hire, the money that would have otherwise been sitting in a bank account, earning interest, has been spent. If the company had to borrow the money to buy the plant or equipment then interest charges are payable on loans and finance agreements.

PRACTICAL TIP

Many construction companies that know they are going to be working on a project for a long period of time will actually buy plant and equipment for that contract. Once the contract has been completed they will sell on the plant and equipment.

Planning the sequence of materials and labour requirements

One of the most important jobs when organising work that will need to be carried out on site is to calculate when, where and how much materials and labour will be needed at any one time. This is organised in a number of different ways. The following headings cover the main documents or processes that are involved.

Bill of quantities

BILL OF QUANTITIES					
Contract			DWG No.		
DESCRIPTION	QUANTITY	UNIT	RATE	AMOUNT	

Figure 2.36 Bill of quantities form

This is used by building contractors when they quote for work on larger projects. It is usually prepared by a quantity surveyor. Fig 2.36 shows you what a bill of quantities looks like.

The form is completed using information from the working drawings, specification and schedule (this is called the take off). It describes each particular job and how many times that job needs to be carried out. It sets the number of units of material or labour, the rate at which they are charged and the total amount.

Programmes of work

A programme of work is also an important document, as it looks at the length of time and the sequence of jobs that will be needed to complete the construction. It has three main sections:

* A master programme that shows the start and finish dates. It shows the duration, sequence and any relationships between jobs across the whole contract.

* A stage programme – this is the next level down and it covers particular stages of the contract. A good example would be the foundation work or the process of making the building weather-tight. Alternatively it might look at a period of up to two months' worth of work in detail.

* A weekly programme – there will be several of these, which aim to predict where and when work will take place across the whole of the site. These are very important as they need to be compared against actual progress. The normal process is to review and update these weekly programmes and then update the stage and master programmes if delays have been encountered.

Stock systems and lead times

One of the greatest sources of delays in construction is not having the right materials and equipment available when it is needed. This means that someone has to work out not only what is needed and how many, but when. It is a balancing act because there are dangers in having the stock on site too early. If all the materials needed for a construction job arrived in the first week then this would cause problems and it is unlikely that there would be anywhere to store them. Materials need to be ordered to ensure that they are on site just before they are needed.

One of the problems is lead times. There is no guarantee that the supplier will have sufficient stock available when you need it. They need to be warned that you will need a certain amount of material at a certain time in advance. This will allow them to either manufacture the stock or get it from their supplier. Specialist materials have longer lead times. These may have to be specially manufactured, or perhaps imported from abroad. All of this takes time.

Once the quantities of materials have been calculated and the sequence of work decided, comparing that to the duration of the project and the schedule, it should be possible to predict when materials will be needed. You will need to liaise with your suppliers as soon as possible to find out the lead times they need to get the materials delivered to the site. This might mean that you will have to order materials out of sequence to the work schedule because some materials need longer lead times than others.

Planning and scheduling using charts

To plan the sequence of materials and labour requirements it is often a good idea to put the information in a format that can be easily read and understood. This is why many companies use charts, graphs and other types of illustrated diagram.

The most common is probably the Gantt chart. This is a series of horizontal bars. Each different task or operation involved in a project is shown on the left-hand side of the chart. Along the top are days, weeks or months. The planner marks the start day, week or month and the projected end day, week or month with a horizontal bar. It shows when tasks start and when they end. It will also be useful in showing when labour will be needed. It will also show which jobs have to be completed before another job can begin. An example of a Gantt chart can be seen in Fig 2.37.

Typical programme for rate of completion on a housing development contract

Figure 2.37 Gantt chart

There are other types of bar chart that can be used to plan and monitor work on the construction site. It is important to remember that each chart relates to the plan of work:

* A single bar – this focuses in on a sequence of tasks and the bar is filled in to show progress.

SINGLE BAR SYSTEM

	ACTIVITY	Week 1	Week 2	Week 3	Week 4
1	Excavate O/site				
2	Excavate Trenches				
3	Concrete Foundations				
4	Brickwork below DPC				

Figure 2.38 Single bar chart

* A two-bar – this tracks the amount of work that has been carried out against the planned amount of work that should have been carried out. In other words it shows the percentage of work that has been completed. It is there to alert the site manager that work may be falling behind and extra resources are needed.

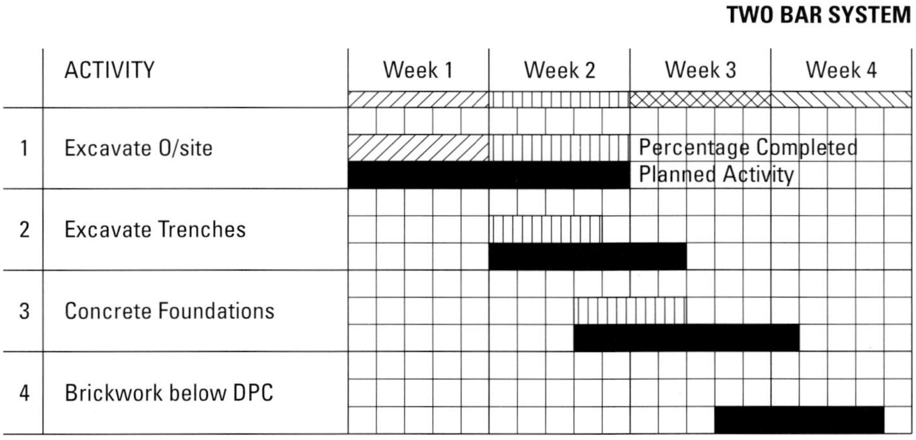

TWO BAR SYSTEM

	ACTIVITY	Week 1	Week 2	Week 3	Week 4
1	Excavate O/site			Percentage Completed	
				Planned Activity	
2	Excavate Trenches				
3	Concrete Foundations				
4	Brickwork below DPC				

Figure 2.39 Two-bar system

* A three-bar – this shows the planned duration of the activity, the actual days that have seen work being done on that activity and the percentage of the activity that has been completed. This gives a snapshot view of how work has progressed over the course of a period of time.

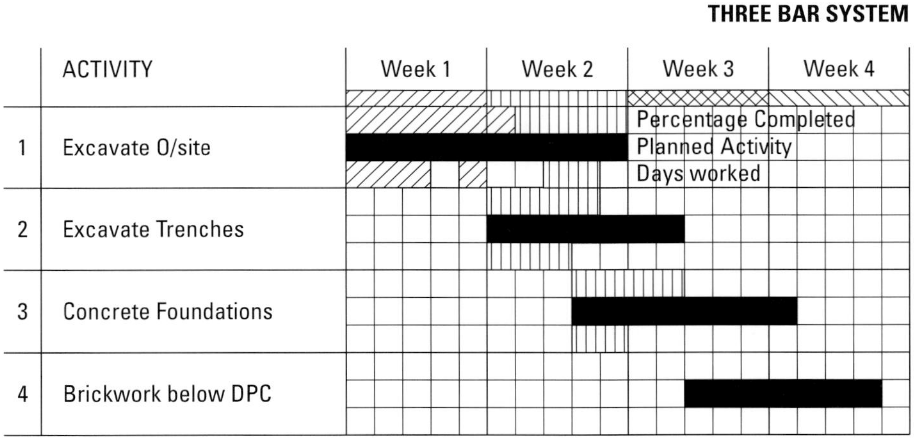

THREE BAR SYSTEM

	ACTIVITY	Week 1	Week 2	Week 3	Week 4
1	Excavate O/site			Percentage Completed	
				Planned Activity	
				Days worked	
2	Excavate Trenches				
3	Concrete Foundations				
4	Brickwork below DPC				

Figure 2.40 Three-bar system

Calculating hours required

We have already seen that different types of construction workers attract different hourly rates of pay. The simple solution in order to work out the cost to complete particular work is to look at the programmes of work and the estimated time required to complete it. The next stage is to estimate how many workers will be needed to carry out that activity and then multiply that by the estimated labour cost per hour.

Added costs

When a construction company estimates the costs of work they have to incorporate a number of other different costs. A summary of these can be seen in the following table.

Added cost	Explanation
National Insurance contributions	Companies employing workers have to pay National Insurance contributions to the government for each employee.
Value Added Tax (VAT)	For businesses that are registered for VAT they have to charge a sales tax on any services that they provide. They collect this money on behalf of the government. The VAT is added to the final cost of the work.
Pay As You Earn (PAYE)	PAYE, or income tax, has to be paid on the income of all workers straight to the government.
Travel expenses	This is particularly relevant if workers on site have to travel a considerable distance in order to do their work. This can reasonably be passed on to the client.
Profit and loss	Many businesses will make the mistake of trying to estimate the costs of their work in the knowledge that they are competing with other businesses, trimming their estimates if they can. As we will see when we look at profitability, a business does need to make money from contracts otherwise it will not have sufficient funds to continue to operate.
Suppliers' terms and conditions	Suppliers will often have set payment terms, such as 30, 60 or 90 days. The construction company needs to have sufficient funds to pay their suppliers on the due date stated on the invoice. If they fail to do this then they could run into difficulties as the supplier may decide not to give them any more credit until outstanding invoices have been paid.
Wastage	It is rarely possible to buy materials and components that are an exact fit. Blocks and bricks, for example, will have a percentage damaged or unusable on each pallet. Materials can also be mis-cut, damaged or otherwise wasted on site. It is sensible for the construction company to factor in a wastage rate of at least 5 per cent. This is often required to allow for cutting and fitting when using stock lengths.
Penalty clauses	Many projects are time sensitive and need to be completed by specified dates. The contracts will state whether there are any penalties to be made if critical dates are missed. Penalty clauses are rather like fines that the construction company pays if they fail to meet deadlines.

Table 2.7

Total estimated prices

As we have seen, the total estimated price needs to incorporate all of the added costs. But there are two other issues that should not be forgotten:

* The cost of any plant and equipment hire – the length of time that these are necessary will have to be calculated together with an additional period in case of delays.

* Contingencies – it is not always possible to determine exact prices, especially in groundworks, and, in any case, not all construction jobs will run smoothly. It is therefore sensible to set aside funds should additional work be needed. This may be particularly true if additional work needs to be done to secure the foundations or if workers and equipment are on site yet it is not possible to work due to poor weather conditions.

Profitability

Setting an initial price for a business's services is, perhaps, one of the most difficult tasks. It needs to take account of the costs that are incurred by the business. It also needs to consider the prices charged by key competitors, as the optimum price that a business wishes to charge may not be possible if competitors are charging considerably lower prices.

It is difficult for a new business to set prices, because its services may not be well known. Its costs may be comparatively higher because it will not have the advantages of providing services on a large scale. Equally, it cannot charge high prices because neither the company nor its services are established in the market.

The process of calculating revenue is a relatively simple task. Sales revenue or income is equal to the services sold, multiplied by the average selling price. In other words, all a business needs to know is how many services have been sold, or might sell, and the price they will charge.

We have already seen that a business incurs costs that must be paid. Clearly these have a direct impact on the profitability of a business.

We can already see that there is a direct relationship between costs and profit. Costs cut into the revenue generated by the business and reduce its overall profitability.

Gross profit is the difference between a company's revenue and its costs. Businesses will also calculate their operating profit. The operating profit is the business's gross profit minus its **overheads.**

A business may also calculate its pre-tax profits, which are its profits before it pays its taxes. It may have one-off costs, such as the replacement of a piece of plant or equipment. These costs are deducted from the operating profit to give the pre-tax profit.

KEY TERMS

Overheads

– these are expenses that need to be paid by the business regardless of how much work they have on at any one time, such as the rent of builders' yard.

The business will then pay tax on the remainder of its profit. This will leave them their net profit. This is the amount of money that they have actually made over the course of a year or on a particular job.

Profits are an important measure of the success of a business. Like other businesses construction companies do borrow money, but profit is the source of around 60 per cent of the funds that businesses use to help them grow.

Businesses can look for ways to gradually increase their profit. They can look at each type of job they do and work out the most efficient way of doing it. This might mean looking for a particular mix of employees, or buying or hiring particular plant and equipment that will speed up the work.

GOOD WORKING PRACTICES

Like any business, construction relies on a number of factors to make sure that everything runs smoothly and that a company's reputation is maintained.

There needs to be a good working relationship between those who work for the construction company, and other individuals or companies that they regularly deal with, such as the local authority, and professionals such as architects and clients.

This is achieved by making sure that these individuals and organisations continue to have trust and confidence in the company. Any promises or guarantees that are made must be kept.

In the normal course of events communication needs to be clear and straightforward. When there are problems accurate and honest communication can often deal with many of them. It can set aside the possibility of misunderstandings.

Good working relationships

Each construction job will require the services of a team of professionals. They will need to be able to work and communicate effectively with one another. Each has different roles and responsibilities.

Although you probably won't be working with exactly the same people all the time you are on site, you will be working with on-site colleagues every day. These may be people doing the same job as you, as well as people with other roles and responsibilities who you need to work with to ensure that the project runs smoothly.

Working with unskilled operatives

It's important to remember that everyone will have different levels of skill and experience. You, as a skilled or trade operative, are qualified in your trade, or working towards your qualification. Some people will be less experienced than you; for example, unskilled operatives (manual workers) are entry level operatives without any formal training. They may, however, be experienced on sites and will take instructions from the supervisor or site manager. You should be patient with colleagues who are less experienced or skilled than you – after all, everyone has to learn. However, if you see them carrying out unsafe practices, you should tell your supervisor or charge-hand straight away.

Working with skilled employees

You'll also work with people who are more experienced than you. It's a good idea to watch how they work and learn from their example. Show them respect and don't expect to know as much as they do if they have been working for much longer. However, if you see them ignoring safety rules, don't copy them; speak to your supervisor.

Working with professional technicians

You might also work with professional technicians, such as civil engineers or architectural technicians. They will have extensive knowledge in their field but may not know as much as you about bricklaying or carpentry. For your relationship to run smoothly, you should respect each other's knowledge, share your thoughts on any issues, and listen to what each other has to say.

Working with supervisors

Supervisors organise the day-to-day running of the site or a team. Charge-hands supervise a specific trade, such as bricklayers or carpenters. They will be your immediate boss, and you must listen to their instructions and obey any rules they set out. These rules enable the site to be run smoothly and safely so it is in your interest to do what your supervisor says.

Working with the site manager

The site manager or site agent runs the construction site, makes plans to avoid problems and meet deadlines, and ensures all processes are carried out safely. They communicate directly with the client. They are ultimately responsible for everything that goes on at the construction site. Even if you don't communicate with them directly, you should follow the guidance and rules that they have put in place. It's in your interest to do your bit to keep the site safe and efficient.

Working with other professionals

You may also need to work with or communicate with other professionals. For example, a clerk of works is employed by the architect on behalf of a client. They oversee the construction work and ensure that it represents the interests of the client and follows agreed specifications and designs. A contracts manager agrees prices and delivery dates. These professionals will expect you to do the job that has been specified and to draw their attention to anything that will change the plans they are responsible for.

Hierarchical charts

A hierarchy describes the different levels of responsibility, authority and power in a business or organisation. The larger the business the more levels of management it will have. The higher up the management structure the more responsibility each person will have.

Decisions are made at the top and instructions are passed down the hierarchy. It is best to imagine most organisations as like a pyramid. The directors or owners of the business are at the top of that pyramid. A site manager may be part-way down and at the bottom of the pyramid are all the workers who are on site.

Trust and confidence

The trust of colleagues develops as a result of showing that the company is reliable, cooperative and committed to the success or goals of the colleague or client. Trust does not happen automatically but has to be earned through actions. An important part of this is building a positive relationship with colleagues. Over time, trust will develop into confidence. Colleagues will have confidence in the company being able to deliver their promises.

The reputation of a construction company is very important. Criticisms of the company will always do far more damage than the positive benefits of a successfully completed contract. This shows how important it is to get things right the first time and every time.

In an industry where there is so much competition, trust and confidence can mean the difference between getting the contract and being rejected before your bid has even been considered.

As the construction company becomes established they will build up a network of colleagues. If these colleagues have trust and confidence in the business they will recommend the business to others.

Ultimately, earning trust and confidence relies on the business being able to solve any problems with the minimum of fuss and delay. Fair solutions need to be identified. These solutions should not be against the interests of anyone involved.

Accurate communication

Effective communication in all types of work is essential. It needs to be clear and to the point, as well as accurate. Above all it needs to be a two-way process. This means that any communication that you have with anyone must be understood.

In construction work it is essential to keep to deadlines and follow strict instructions and specifications. Failing to communicate will always cause confusion, extra cost and delays. In an industry such as this it is unacceptable and very easy to avoid. Negative communication or poor communication can damage the confidence that others have in you to do your job.

It is important to have a good working relationship with colleagues at work. An important part of this is to communicate in a clear way with them. This helps everyone understand what is going on, what decisions have been made. It also means being clear. Most communication with colleagues will be verbal (spoken). Good communication results in:

* cutting out mistakes and stoppages (saving money)

* avoiding delays

* making sure that the job is done right the first time and every time.

The more complex a contract, the more likely it is that changes and alterations will be needed. The longer the contract runs for, the more likely it is that changes will happen. Examples are as follows:

* Alterations to drawings – this can happen as a result of several different factors. The architect or the client may decide at a fairly late stage that changes need to be made to the design of the project. This will require all documents that rely on information from the drawings to be amended. This could mean changes to the schedule, specification and work programmes and the need for materials and labour at particular times.

* Variation to contracts – although the construction company may have agreed with the client to carry out work based on particular drawings and specifications, changes to design and to the requirements may happen. It may be necessary to put in new estimates for additional work and to inform the client of any likely delays.

* Changes to risk assessments – it is not always possible to predict exactly what hazards will be encountered during a project. Neither is it possible to predict whether new legislation will come into force that requires extra risk assessments.

* Work restrictions – although the site will have been surveyed for access and cleared of obstacles such as low trees, problems may arise during the work. Local residents, for example, may complain about lighting and noise. This could reduce working hours on site. This could all have an impact on the schedule of work.

* Change in circumstances – this could cover a wide variety of different problems. Key suppliers may not be able to deliver materials or components on time. Tried and trusted sub-contractors may not be available. The client may run out of money or a problem may be unearthed during excavation and preparation of the site.

REED TIP

Open and frank communication means being able to say no if something is not possible. It's OK to say that something can't be done, rather than saying 'yes, yes, yes' and then being unable to complete a task.

DID YOU KNOW?

During the construction of the Olympic basketball site in London the whole site had to be evacuated when a Second World War bomb was found. It had to be removed by specialists before work could continue.

TEST YOURSELF

1. What is the system that is gradually taking over from CAD as the main way to produce construction drawings?
   a. CTIBM
   b. CIM
   c. BIM
   d. SIM

2. What does a block plan show?
   a. The construction site and its surrounding area
   b. Local boundaries and roads
   c. Elements and components
   d. Constructional details

3. Who might give you a delivery note?
   a. A postal worker
   b. An architect
   c. A contractor
   d. A supplier

4. Which type of construction drawings show the different faces or views of an object?
   a. Orthographic
   b. Section
   c. Elevation
   d. Plan

5. How many different types of pictorial projection are there?
   a. 3
   b. 4
   c. 5
   d. 6

6. Which of the following is usually a bid for a fixed amount of work in competition with other companies?
   a. Estimate
   b. Quotation
   c. Tender
   d. Invoice

7. To calculate the area of a room, which two measurements are needed?
   a. Length and height
   b. Height and width
   c. Length and width
   d. Length and circumference

8. If you had 5 workmen being paid £25 per hour and they were working for 4 hours, what would be the total labour cost?
   a. £125
   b. £250
   c. £500
   d. £600

9. Some contracts state that if a deadline is missed a fine has to be paid. What are these called?
   a. Terms and conditions
   b. Wastage
   c. Penalty clause
   d. Critical date payment

10. A business's operating profit is its gross profit minus which of the following?
    a. Tax
    b. Overheads
    c. Net profit
    d. Labour costs

Chapter 3

Unit CSA–L3Core08
ANALYSING THE CONSTRUCTION INDUSTRY AND BUILT ENVIRONMENT

LEARNING OUTCOMES

LO1: Understand the different activities undertaken within the construction industry and built environment

LO2: Understand the different roles and responsibilities undertaken within the construction industry and built environment

LO3: Understand the physical and environmental factors when undertaking a construction project

LO4: Understand how construction projects can benefit the built environment

LO5: Understand the principles of sustainability within the construction industry and built environment

INTRODUCTION

The aim of this unit is to:

* help you understand more about the construction industry and its place in society.

CONSTRUCTION INDUSTRY AND BUILT ENVIRONMENT ACTIVITIES

Half of all the non-renewable resources used across the globe are consumed by construction. Construction and the built environment are also linked with the pollution of drinking water, the production of waste and poor air quality.

Nevertherless, buildings create wealth. In the UK, buildings represent three-quarters of all wealth. Buildings are long-term assets. Today it is recognised that buildings should have the ability to satisfy user needs for extended periods of time. They must be able to cope with any changing environmental conditions. They also need to be capable of being adapted over time as designs and demands change.

There is an increasing move towards naturally lit and well-ventilated buildings. There is also a move towards buildings that use alternative energy sources.

The first part of this chapter looks at the type of work that has developed around the broader construction industry and built environment. It looks at the work that is undertaken and the different types of clients who use the construction industry.

Range of activities

The construction industry and the broader built environment is a highly complex network of different activities. While there are a great many small businesses that focus on one particular aspect of construction, they need to be seen as part of a far larger industry. Increasingly it is a global industry, with major business organisations operating not just in the UK but also in a wide variety of locations around the world. Their skills and expertise are in great demand wherever there is construction. The following table outlines some of the activities that are undertaken by the construction industry and the broader built environment.

Activity	Description
Building	This is the accepted and traditional activity of the construction industry. It involves building homes and other structures, from garden walls to entire housing estates or even Olympic villages.
Finishing	Finishing refers to a part of the industry that focuses on decorative work, such as painting and decorating. Once buildings are completed, in order to make them ready for habitation a broad range of professions are needed. Plumbers will install water and sanitation. Electricians will connect electrical services and equipment. Interior designers will create the desired look for the building.
Architecture	Architects and technicians design buildings for clients. The structures are designed to meet the needs of the client while ensuring that they conform to Building Regulations, local planning laws and decisions, as well as other legislation such as CDM Regulations and ensure they are sustainable.
Town planning	Town planning involves organising the broader built environment in a particular area. Town planners need to examine each planning application and see how it fits into the overall long-term future of the area. They need to ensure that the area meets the needs of future generations.
Surveying	This involves measuring and examining land on which building or other external work will take place. It can involve setting out the building. Surveyors use drawings by an architect to correctly position the building. They will be able to work out the area of the building and any volumes. Building surveyors check that the building is structurally sound, while quantity surveyors look after costs.
Civil engineering	Civil engineers are usually involved in major projects, such as road and railway building, the construction of dams, reservoirs and other projects that are not usually buildings. They are involved in what is known as infrastructure projects, such as transport links, networks and hubs.
Repair and maintenance	All buildings need professionals who are able to repair and maintain a broad range of features. From the foundations to the roof, carpenters, builders, electricians, plumbers and more specialist companies, such as pest control, can all be considered to be part of the repair and maintenance side of the industry. Pre-1919 buildings also have particular requirements and are maintained and repaired in a way that suits their construction and to avoid further damage or inappropriate work that will look out of place. This requires people who have specialist heritage skills.
Building engineering services	When buildings are occupied they need to be continually supported in terms of a rigorous checking and maintenance programme. This part of the industry can deal with lifts and escalators, lighting and heating, fire alarms and other inbuilt systems.
Facilities management	For larger commercial buildings or hospitals, schools, colleges and universities, systems need to be in place to replace parts of the building if they wear out or are damaged. This includes cleaning, air conditioning companies, painting and decorating, replacement of doors, windows and a host of other activities.
Construction site management	Construction sites can be complex and demanding places and someone needs to organise them and to monitor progress. Construction site management involves organising the delivery of materials, security, safety, the management of the workforce and contractors.
Plant maintenance and operation	Just as commercial buildings and dwellings need constant maintenance, so too do factories and other sites where products are made or processes are carried out. These individuals can be involved in the energy industry, at gas, oil and nuclear plants, or be responsible for maintaining factories that produce vehicles or food.
Demolition	Demolition experts are responsible for levelling sites in a safe and controlled way. They may have to demolish buildings that could contain asbestos or they may have to use controlled explosions.

Table 3.1

Types of work

There is also a wide range of work that is undertaken in different sectors. Some of this is very specialised work. Some companies will focus purely on that type of work, gaining a reputation and expertise in that area. The following table outlines the types of work that is undertaken within the construction industry.

Type of work	Description
Residential	This is any work connected with domestic housing or dwellings. It can include the building of new homes, extensions or renovations on existing homes and the construction of affordable accommodation for organisations such as housing associations.
Commercial	This is work related to any buildings used by businesses. It can include factories, office blocks, production units, industrial units or private hospitals.
Industrial	This is more specialist work, as it can involve construction, including civil engineering, of heavy industrial factories, such as oil refineries or plants for car manufacturing.
Retail	This can include building or refurbishing shops in high streets or the construction of out-of-town retail parks.
Recreational and leisure	Many of these projects are designed for use by communities, such as sports facilities, fitness clubs, leisure centres, swimming pools and other community sports projects. In the past decade the construction industry was involved in the various London 2012 Olympic facilities.
Health	This includes specialist building services to create hospitals and other health facilities, such as doctors' surgeries and care homes.
Transport infrastructure	This is another broad area of work that includes roads, motorways, bridges, railways, underground trains and tram systems, as well as airports, bus routes and cycle paths.
Public buildings	This is the building and maintenance of large buildings for local and central government. It can include offices, town halls, art galleries, museums and libraries.
Heritage	Heritage involves work on listed properties of historical importance. This is a specialist area, as Building Regulations, planning laws and Listed Status require any work to be carried out in sympathy with the original design of the building.
Conservation	This is an increasingly important area of work, as it involves the protection of natural habitats. It would involve work in National Parks, Areas of Outstanding Natural Beauty, animal sanctuaries and could also include construction work related to coastal erosion and flood defences.
Educational	This is the construction of schools, colleges, universities and other buildings used for educational purposes.
Utilities and services	This is work that is related to the installation, maintenance and repair of the key utilities, which include gas, electricity and water.

Table 3.2

Types of client

As we have seen, there is a huge range of activities and types of work in the broader construction industry and built environment areas. This means that there is a huge range of different potential types of client. Some are private individuals but at the other end of the scale they might be huge companies or government departments. The following table outlines the range of different types of client.

Type of client	Description
Private	These are usually individual owners of homes or buildings. They may be people who want work done on their own homes, or on their own business premises, such as a small shop. Many of the individual shop or business owners may be sole traders. These are individuals who run and own a small business.
Corporate	Corporate is a term that is used to describe larger companies or businesses. They can be individuals that run factories, larger shops, industrial units or some kind of service-based organisation, including banks, insurance companies and estate agents. Some construction companies have long-term contracts with corporate businesses, which have many branches around the country. There is a rolling programme of maintenance, upgrading and repair. The companies can be public limited companies (PLC), who are owned by shareholders with their shares traded on the Stock Exchange.
Government	The government can be a client on a local, regional or national basis. This part of the industry has become more complicated, as there are multiple levels of government across the UK. There is also a Scottish Parliament and a Welsh Assembly, in addition to the UK Parliament based in London. Local councils will be responsible for maintaining a wide range of services and they will also be involved in construction. This includes schools, roads, the maintenance of social housing and parks and leisure facilities. In addition to this there are government departments based in London with regional offices, such as the Ministry of Defence, which is responsible for facilities related to the armed forces, and the National Health Service, which is responsible for hospitals and other health provision. The government (including local authorities and non-departmental public bodies) must comply with strict procurement (buying) rules, which often involve tenders. They also have limited budgets, which could affect the building project's schedule.

Table 3.3

CONSTRUCTION INDUSTRY AND BUILT ENVIRONMENT ROLES AND RESPONSIBILITIES

As we have discussed, the construction industry and the built environment is a complex network of different activities. As the industry has developed over time it has become important for individuals to specialise and take on specific roles and responsibilities.

Roles and responsibilities of the construction workforce

The following tables show the broad range of different roles and briefly outline their responsibilities within the construction industry.

From the design and planning phase onwards

Role	Responsibilities
Client	The client, such as a local authority, commissions the job. They define the scope of the work and agree on the timescale and schedule of payments.
Customer	For domestic dwellings, the customer may be the same as the client, but for larger projects a customer may be the end user of the building, such as a tenant renting local authority housing or a business renting an office. These individuals are most affected by any work on site. They should be considered and informed with a view to them suffering as little disruption as possible.
Architect	They are involved in designing new buildings, extensions and alterations. They work closely with clients and customers to ensure the designs match their needs. They also work closely with other construction professionals, such as surveyors and engineers.
Estimator	Estimators calculate detailed cost breakdowns of work based on specifications provided by the architect and main contractor. They work out the quantity and costs of all building materials, plant required and labour costs.
Planner	Consultant planners such as civil engineers work with clients to plan, manage, design or supervise construction projects. There are many different types of consultant, all with particular specialisms.
Buyer	This individual works closely with the quantity surveyor. It is the buyer's job to source suitable materials as specified by the architect. They will negotiate prices and delivery dates with a range of suppliers.

Table 3.4

Surveying

Role	Responsibilities
Land agent	This is an individual who is authorised to act as an agent in the sale of land or buildings by the owner. Basically they are estate agents that sell plots of land.
Land surveyor	A land surveyor measures, records and then produces a drawing of the landscape. The data that they produce is used to plan out construction work.
Building surveyor	A building surveyor is responsible for making sure that both old and new buildings are structurally sound. They are involved in the design, maintenance, repair, alteration and refurbishment of buildings.
Quantity surveyor	Quantity surveyors are concerned with building costs. They balance maintaining standards and quality against minimising the costs of any project. They need to make choices in line with Building Regulations. They may work either for the client or for the contractor.

Table 3.5

Engineering

Role	Responsibilities
Building services engineers	They are involved in the design, installation and maintenance of heating, water, electrics, lighting, gas and communications. They work either for the main contractor or the architect and give instruction to building services operatives.
Structural engineer	Structural engineers are involved in ensuring that construction work is strong enough to deal with its use and the external environment. So they will be involved in the shape, design and the materials used. They will not only deal with new construction work but also advise on older buildings or buildings that have been damaged.
Consulting/building engineer	These individuals are involved in site investigation, building inspection and surveys. They get involved in a wide range of construction and maintenance projects.
Plant engineer	A plant engineer is responsible for maintaining and repairing a variety of machinery and equipment. They will also install and modify machinery and equipment in factories as part of an industrial or manufacturing process.
Site engineer	A site engineer is involved in setting out the plans for sewers, drains, roads and other services.
Specialist engineer	A good example of a specialist engineer is one that deals entirely with insulation. They will advise and install a range of energy conservation materials and equipment. A geotechnical engineer is another example. They carry out investigations into below foundation level and look at rock, soil and water.
Mechanical engineer	Mechanical engineers are primarily involved in installing and maintaining machinery and tools. It is a wide ranging profession but they will have overall responsibility for their particular area of work.
Demolition engineer	These engineers perform the task of tearing down old structures or levelling ground to make way for new buildings.
Infrastructure engineer	These engineers deal with the planning, construction and management of roads, bridges and similar structures.

Table 3.6

REED TIP

Any work experience is relevant to your job applications. It doesn't have to be paid work – e.g. volunteering to help run Scout and Guide activities shows your sense of responsibility. Think of the times when others have had to rely on you.

CASE STUDY

South Tyneside Homes

South Tyneside Council's Housing Company

Your apprenticeship is just the start

Gary Kirsop, Head of Property Services, started at South Tyneside Homes as an apprentice 24 years ago.

'After becoming qualified, I had two options. I could have stayed working on the sites and become a site manager or technical assistant. I qualified as a building surveyor, doing my advanced craft at Sunderland College. After that, I went to Newcastle College to do my ONC and CHND, and eventually went on to finish a degree at Newcastle.

When I was a technical assistant I worked on education and public buildings, and spent a year in housing. As a technical assistant I was working on drawing (CAD), estimating small jobs to large jobs. Then an opportunity for Assistant Contracts Manager on capital works came up. Since then, I've also worked in disrepair and litigation, as well as two years with the empty homes department, and I've worked as a Construction Services Manager, responsible for the capital side, new homes, decent homes, and the gas team.

Four years ago, the Head of Property Services job came up and it's been a fantastic opportunity – my team has been one of the best in the country for performance. My department is responsible for repairs and maintenance, capital works, empty homes, and management of the operational side. We do responsive repairs for emergency situations, planned repairs, work for the "Decent Homes" programme where we bring properties up to standard, and we've recently built four new bungalows. Anything in construction, we have the skills and labour to do it in property services.

The full management team here in property services all started as apprentices, like me. It really helps that we understand the whole process from beginning to end.

So you can see that doing your apprenticeship is not only great in itself, but it also gives you skills for life and ongoing opportunities for education, training and your career.'

PHYSICAL AND ENVIRONMENTAL FACTORS AND CONSTRUCTION PROJECTS

Increasingly, people working in construction and the built environment are being asked to ensure that they minimise physical and environmental impacts when carrying out construction work. Construction has an enormous impact on the environment. Environmental measures will depend on the nature of the work and the site. For example, excavations that result in changes in the levels of land can cause problems with water quality and soil erosion. Many of these negative impacts can be reduced during the planning stage.

Physical and environmental factors

Physical factors relate to the impact that any new construction project will have on any existing structures and their occupants. Any new construction project is going to have a negative impact on home owners and businesses. There will be increased traffic on roads and a host of other considerations.

Once the construction has been completed there may be longer term impacts. A prime example would be building a new housing development in an area that lacks good roads, sufficient schools or access to health facilities. During the planning and development stage these factors will be looked at to see what the knock-on effects might be in the short and long term.

Environmental factors concern the impact that a construction project has on the natural environment. This would include any possible impacts on trees and vegetation, wildlife and habitats. It can also have an impact on the air quality or noise levels in the area.

Physical factors and the planning process

There is a wide variety of different physical factors that have to be taken into consideration during the planning process. These are outlined in the following table.

Physical factor	Explanation
Planning requirements	The majority of new developments or changes to existing buildings do require consent or planning permission. The local planning authority will make a decision whether any such construction will go ahead. Each authority has a development framework that outlines how planning is managed. This includes the change of use of a building or a piece of land.
Building Regulations	There are 14 technical parts of the Building Regulations covering everything from structural safety to electrical safety. They also outline standards of quality of work and materials used. All new developments and major changes to existing buildings must comply with Building Regulations.
Development or land restrictions	This is a complicated area, as there are often many restrictions on building and the use of land. One of the most complex is restrictive covenants, which are created in order to protect the interests of neighbours. They might restrict the use of the land and the amount of building work that can take place.
Building design and footprint	The footprint is the physical amount of space or area that the proposed development takes up on a given plot of land. There may be limits as to the size of this footprint. In terms of building design, certain areas may have restrictions as the local authority may not approve the construction of a building that is out of character, or that would adversely affect the overall look of the area.
Use of building or structure	Each building or structure will have a Use Class, such as 'residential', 'shops' or 'businesses'. Redeveloping an existing building and not changing the use to which it is put, for example renovating a building from a butcher to a chemist, does not usually require planning permission. However, changing from a bank to a bar would require planning permission. Certain uses, due to their unique nature, do not fall into any particular Use Class and planning permission is always required. A good example would be a nightclub or a casino.

Physical factor	Explanation
Impact on local amenities	During the construction phase it is likely that roads or access may have to be blocked, which could impact on local businesses. In the longer term additional traffic and the need for parking may have an impact on local amenities, as will the demand for their use.
Impact on existing services and utilities	Any new development or major change in use of an existing structure may put extra strain on services and utilities in the area. A new housing development, for example, would require power cables to be run to the site. It would also need excavation work to connect it to the sewers and underground pipes run onto the site for potable water. All of this is potentially disruptive and may require considerable investment by the utility or service provider.
Impact on transportation infrastructure	Major new developments will have a huge impact on the roads and public transport in an area. Permission for major developments often comes with the requirement to improve access routes, build new roads and the requirement to make a contribution to improvements in the infrastructure. New developments can radically change the flow of traffic in an area and may have a knock-on effect in terms of maintenance and repair in the longer term.
Topography of the proposed development site	The term topography refers to the location of the site and how dominant it will be in the local landscape. Obviously a development that is situated on a hill or ridge is far more obvious and will have a longer lasting impact on the local area. If the development is considered to be too obtrusive or visible then it may be deemed as inappropriate to situate the development on that site.
Greenfield or brownfield site	A greenfield site is an area of land that has never been used for non-agricultural purposes. A brownfield site is usually former industrial land, or land that has been used for some other purpose and is no longer in use. There is more information on greenfield and brownfield sites in the next section of this chapter.

Table 3.7

Environmental factors and the planning process

Just as there are physical factors, there are also different environmental factors that need to be considered. Some of the major ones are detailed in the following table.

Environmental factor	Explanation
Topography of the development site	As mentioned in the previous table, the topography of the development site can have a marked impact on the local environment. It may dominate what is otherwise a predominantly natural environment, perhaps with woodland or rolling hills.
Existing trees and vegetation	Sites may have to be cleared in order to provide the necessary space for the footprint of the structure. It may be prohibited to remove or otherwise interfere with certain trees and vegetation, as they may be protected. The normal course of events is to minimise the impact on existing plant life and to have a replanting phase after the site has been developed.
Impact on existing wildlife and habitats	Any potential impact on wildlife and plants that are under threat could mean that the site would not receive the go ahead. An environmental impact study will identify whether there are any specific dangers that will affect the natural habitat of the area, or endanger any local species of wildlife.
Size of land and building footprint	There is a formula that determines the usually permitted footprint of a piece of land compared to the actual size of the plot of land. For example, a 4-bedroom house on an average housing estate would take up approximately 1/12th of an acre (11.5 m × 29 m).

Environmental factor	Explanation
Access to the building or structure	It is not only the building plot that needs to be considered in terms of its environmental impact. Access to the site is another concern. Existing roads may have to be widened, perhaps a roundabout installed. Alternatively new roads may have to be built across other plots of land. For pedestrian traffic footpaths may also be necessary. These can either be alongside existing roads or built alongside new roads, requiring even more space. There may be existing footpaths and this could mean that access needs to be provided through the site or the footpaths diverted.
Supply of services to the building or structure	Running above ground services and utilities to the site may also present a problem as far as its impact on the environment is concerned. It may not be possible to allow features such as pylons or street lights to dominate the landscape.
Natural water resources	New developments can affect the biodiversity of an area by impacting on natural waterways. Local wildlife and plants rely on this resource. In addition to this, construction could either pollute or affect the quality of the local water.
Land restrictions	There may be land restrictions that limit either the use or the size of any development. Developments will not be allowed to adversely affect surrounding properties and owners. There are conservation areas, scheduled monuments, archaeological sites and scheduled or listed buildings. These are all protected and construction on or near them is either prohibited or severely limited.
Future development and expansion	Although the intention may be to restrict the environmental impact of the site in the first phase of development, in the future this might not be possible. Major housing development is often carried out in phases and the size of the development will gradually increase as demand increases. It is therefore important when permission is initially given that the likelihood of future development and expansion is taken into account.

Table 3.8

Figure 3.1 Trees on a proposed site may need to be protected during construction work

DID YOU KNOW?

In some cases, Tree Preservation Orders are put in place by the local planning authority. These prevent the removal of trees or work on them without permission. Some land, due to its natural beauty, importance to local wildlife and plants or special geological features, can also be protected, making it impossible for any development to take place on the site.

HOW CONSTRUCTION PROJECTS BENEFIT THE BUILT ENVIRONMENT

The construction industry is one of the UK's largest employers. It is a hugely diverse industry. Construction projects can have a massive impact on the built environment. They can rejuvenate whole areas; improve the housing stock, amenities and the general life and well-being of the local population. The built environment describes the overall look and layout of a specific area. Each new construction project and its architectural design will have an impact on that built environment and the broader, natural environment. If it is carefully and sympathetically planned and organised it can have a positive impact on the way people live, work and interact with one another.

Each new development has enormous environmental, social and economic consequences. Increasingly it has a role to play in ensuring that our built environment has a strong and sustainable future.

Land types available for development and their advantages and disadvantages

In March 2012 the National Planning Policy Framework was published, which aims to review planning guidance across the UK. The idea was to encourage the building of domestic dwellings. It stated that there would be a policy to try to use as many brownfield sites as possible, but that greenfield sites in rural areas would no longer be protected at any cost. Where development was necessary it would take place, as there was a huge demand for homes, shops and workplaces.

The first targets for development would be sites that had been used in the past for other purposes.

Greenfield land or sites

Greenfield sites are usually either agricultural or amenity land. Given the fact that there is a housing crisis in the UK and that land needs to be allocated to build millions of new homes, greenfield sites are very much under consideration.

The problem in doing this is that there is huge resistance, particularly in rural areas, to losing greenfield sites for the following reasons:

* Once a greenfield site has been developed it is extremely unlikely that it will ever return to agricultural use. Any loss of agricultural land means a reduction in the amount of food that can be produced in the UK. There might also be a drop in employment in the local area as fewer farm workers are needed.

* Natural habitats of wildlife and plants are destroyed forever.

* Greenfield or amenity land, if lost, means that the land can no longer be used for leisure and recreation.

* Developments on greenfield sites can have a negative impact on the local transport infrastructure and will increase the amount of energy used because things are further away from town centres.

* The loss of green belts of agricultural land around cities, towns and villages means that each separate area loses its identity and in effect becomes a suburb of a larger town or city.

DID YOU KNOW?

London's boundaries may be expanded by 40 miles (65km), taking London into parts of Kent, Essex and Bedfordshire. In this new area alone 200,000 new homes could be built on both green and brownfield sites. It would create some 300,000 new jobs. (Source: www. sustainablebuild.co.uk)

Figure 3.2 Building on greenfield and greenbelt land is a controversial issue

Brownfield land or sites

Brownfield sites are pieces of land that have been previously developed. They were probably used for either industrial or commercial purposes, but are now derelict and abandoned.

Figure 3.3 Brownfield sites have already been built on

Brownfield sites can be found in areas where there is a high demand for new homes. It has been estimated that there are more than 66,000 hectares of brownfield sites in England alone. At least a third of this land can be found in the southeast of England, where there is the highest demand for housing. Around 60 per cent of new housing is being built on brownfield sites. This is a trend that is likely to accelerate over the next 10 years.

Brownfield sites are not just used for housing projects but are also sites for commercial buildings, as well as recreational sites and newly planted woodland.

Reclaimed land

There are areas, particularly around the coast and in estuaries, which for many years have been bogs or salt marshes. These damp grasslands can be gradually drained of water and eventually provide agricultural land or, in some cases, land suitable for housing developments. With global warming and climate change threatening to permanently flood huge areas of the UK, it may seem strange to consider humans reversing the process.

The area is converted by digging flood relief channels and drainage ditches to encourage the water to flow out and away from the land. To protect the land during this process banks are built to keep out river and seawater. It is a long and involved process but can provide possible land for redevelopment. This process has been successful in many different parts of the world, notably in the Fens in East Anglia, on the Netherlands coast, where pumping stations reclaim land from the sea, and in the Middle and Far East where huge projects have reclaimed vast areas of land.

DID YOU KNOW?

Singapore is a huge, vibrant island city. But 200 years ago virtually nobody lived there and it was a swamp. By 2030 it is predicted that at least 150km^2 of new land will be made available.

Figure 3.4 Reclaiming land enables it to be put other uses

Contaminated land

Many brownfield sites, particularly those once used for industrial purposes, are contaminated with varying levels of hazardous waste and pollutants. Before any development can take place an environmental consultant will organise the analysis of soil, ground water and surface water to identify any risks.

Special licences are required to reclaim brownfield sites and this can be a very expensive process for developers. The main way of dealing with brownfield sites is a process known as remediation. This involves the removal of any known contaminants to a level that will not affect the health of anyone living or working on the site both during construction and after building is complete.

Not all brownfield sites are, therefore, suitable or cost-effective. In some cases the cost of removing the contaminants exceeds the value of the land after it has been developed. There are new ways of dealing with contaminants:

* Bioremediation – this uses bacteria, plants, fungi and micro-organisms to destroy or neutralise contaminants.

* Phytoremediation – plants are encouraged to grow on the site and the contaminants are taken up into the plant and stored in their leaves and stems.

* Chemical oxidation – this involves injecting oxygen or oxidants into contaminated soil and water to destroy contaminants.

DID YOU KNOW?

Brownfield redevelopment has huge advantages as it not only deals with environmental health hazards, but also regenerates areas. It can provide affordable housing, jobs and conservation.

Figure 3.5 Contaminated land must be cleaned before use

Social benefits of construction development

The construction industry and the built environment do provide a range of potential benefits, particularly to local areas. These are examined in the following table.

Social benefit	Explanation
Regeneration of brownfield sites	Disused land, usually former industrial sites, and have been developed for new housing and commercial sites. In London, virtually the whole of the 2012 Olympic village was built on brownfield sites.
Local employment	Construction sites need the skills of local construction workers and offer opportunities for small businesses. Long-term projects offer long-term employment for local people.
Improved housing	New developments and refurbishment of older properties provide greener and more energy efficient dwellings. This has a long-term positive impact for the environment and the reduction in the use of non-renewable resources.
Improvements to local infrastructure	A new development of any size often comes with the requirement for the developers to contribute towards the building of new roads and other infrastructure projects for the area. New developments, in order to work, need access roads, transport and other facilities.
Improvements to local amenities	Modern housing developments and commercial properties need to have amenities near them in order to make them viable in the longer term. This means the building of schools, hospitals, health centres and shops.

Table 3.9

Figure 3.6 Sustainable developments aim to be pleasant places to live

SUSTAINABILITY

Carbon is present in all fossil fuels, such as coal or natural gas. Burning fossil fuels releases carbon dioxide, which is a greenhouse gas linked to climate change.

Energy conservation aims to reduce the amount of carbon dioxide in the atmosphere. The idea is to do this by making buildings better insulated and, at the same time, making heating appliances more efficient. It also means attempting to generate energy using renewable and/or low or zero carbon methods.

According to the government's Environment Agency, sustainable construction is all about using resources in the most efficient way. It also means cutting down on waste on site and reducing the amount of materials that have to be disposed of and put into **landfill.**

In order to achieve sustainable construction the Environment Agency recommends:

* reducing construction, demolition and excavation waste that needs to go to landfill

* cutting back on carbon emissions from construction transport and machinery

* responsibly sourcing materials

* cutting back on the amount of water that is wasted

* making sure construction does not have an impact on **biodiversity.**

What is meant by sustainability?

In the past buildings have been constructed as quickly as possible and at the lowest cost. More recently the idea of sustainable construction has focused on ensuring that the building is not only of good quality and that it is affordable, but that it is also energy efficient.

Sustainable construction also means having the least negative environmental impact. So this means minimising the use of raw materials, energy, land and water. This is not only during the build period but also for the lifetime of the building.

KEY TERMS

Landfill

– 170 million tonnes of waste from homes and businesses are generated in England and Wales each year. Much of this has to be taken to a site to be buried.

Biodiversity

– wherever there is construction there is a danger that the wildlife and plants could be disturbed or destroyed. Protecting biodiversity ensures that at risk species are conserved.

Figure 3.7 Eco houses are becoming more common

Construction and the environment

In 2010, construction, demolition and excavation produced 20 million tonnes of waste that had to go into landfill. The construction industry is also responsible for most illegal fly tipping (illegally dumping waste). In any year there are at least 350 serious pollution incidents caused as a result of construction.

Figure 3.8 Always dispose of waste responsibly

KEY TERMS

Carbon footprint

– this is the amount of carbon dioxide produced by a project. This not only includes burning carbon-based fuels such as petrol, gas, oil or coal, but includes the carbon that is generated in the production of materials and equipment.

DID YOU KNOW?

Search on the internet for 'sustainable building' and 'improving energy efficiency' to find out more about the latest technologies and products.

Regardless of the size of the construction job, everyone in construction is responsible for the impact they have on the environment. Good site layout, planning and management can help reduce this impact.

Sustainable construction helps to encourage this because it means managing resources in a more efficient way, reducing waste and reducing your **carbon footprint.**

Finite and renewable resources

We all know that resources such as coal and oil will eventually run out. These are examples of finite resources.

Oil is not just used as fuel – it is used in plastic, dyes, lubricants and textiles. All of these are used in the construction process.

Renewable resources are those that can be produced by moving water, the sun or the wind. Materials that come from plants, such as biodiesel, or the oils used to make some pressure-sensitive adhesives, are examples of renewable resources.

The construction process itself is only part of the problem. It is also the longer term impact and demands that the building will have on the environment. This is why there has been a drive towards sustainable homes and there is a Code for Sustainable Homes, which is a certification of sustainability for new build housing.

The future

Sustainability also means ensuring that future generations do not suffer from the ill-considered activities of today's generation. The following table outlines some of the present dangers and concerns.

Present or future concern	Explanation
Global shortages	Many naturally found resources will eventually run out and they will have to be replaced with alternatives. Acting now to discover, develop or use alternatives will delay this. Construction is at the forefront of finding alternatives and looking at different construction materials and methods.
Needs of future generations	Buildings constructed today must to be useful and affordable for future generations. At the same time, materials and construction methods should not leave a bad legacy that future generations have to deal with.
Global warming	The construction industry has been criticised over its contribution to global warming. A lack of co-ordination between different parts of the industry has produced poor quality, energy-inefficient buildings. The government is keen to ensure that the industry trains people about the principles of sustainable design and efficient technologies. These steps need to be put in place to inform decisions at the design stage of a building.
Climate change	Construction projects need to take into account the effects of climate change and consider ways to reduce the project's impact on the environment. This means minimising carbon emissions, using sustainable (or renewable) energy and reducing water consumption.
Extinction of species and vegetation	Global warming and climate change has an impact on animals and plants. On a local level, this is also a problem as construction can destroy natural habitats. Increasingly, this is closely monitored and environmental impact studies are used to prevent this from happening.
Destruction of natural resources	There are strict planning laws that aim to prevent the industry from destroying or harming natural habitats. Ancient woodland, sites of scientific importance and other sites of interest are all protected. It is also the case that development in areas that are likely to flood or cause flooding are prohibited or controlled.

Table 3.10

KEY TERMS

Global warming

– a rise in temperature of the earth's atmosphere. The planet is naturally warmed by rays, some being reflected back out into space. The atmosphere is made up of gases (some are called greenhouse gases) which are mainly natural and form a kind of thermal blanket. The human-made gases are believed to make this blanket thicker, so less of the heat escapes back into space. Over the past 100 years, our climate has seen some rapid changes. This is believed to be linked to changes in the makeup of the atmosphere and land use.

Climate change

– the burning of fossil fuels (coal, gas, wood, oil) has resulted in an increase in the amount of greenhouse gases. This has pushed up global temperatures. Across the world, millions do not have enough water, species are dying out and sea levels are rising. In the UK we see extreme events such as flooding, storms, sea level rise and droughts. We have wetter warmer winters and hotter drier summers.

Figure 3.9 Climate change may be a serious problem over the next decades

Social regeneration

Construction projects are often used to regenerate areas of the UK that have lacked investment in the past. As industry develops and changes over time, whole areas that would once have been extremely busy in the past now have empty industrial units and high unemployment levels. As the area loses jobs housing deteriorates, as does the local infrastructure, as there is no money in the local economy.

Redeveloping these waste sites is seen as a way in which a whole area can be regenerated or reborn. Construction projects bring jobs relating to the project but they also bring the promise of longer term jobs. These areas have relatively cheap land and lower rents. Also the workforce expects lower rates of pay. This attracts businesses to relocate to the new buildings created by construction developers. This brings work, improved housing, and improvements to the local infrastructure and amenities.

Sustainability and its benefits

Energy efficiency is all about using less energy to provide the same result. The plan is to try to cut the world's energy needs by 30 per cent before 2050. This means producing more energy efficient buildings. It also means using energy efficient methods to produce the materials and resources needed to construct buildings.

Alternative methods of building

The most common type of construction in the UK is brick and blockwork. However there are plenty of other options:

* timber frame – using pre-fabricated timber frames which are then clad

- insulated concrete formwork – where a polystyrene mould is filled with reinforced concrete

- structural insulated panels – where buildings are made up of rigid building boards rather like huge sandwiches

- modular construction – this uses similar materials and techniques to standard construction, but the units are built off site and transported ready-constructed to their location.

Figure 3.10 Insulated concrete formwork

Figure 3.11 Modular construction

There are alternatives to traditional flooring and roofing, all of which are greener and more sustainable. Green roofing (both living roofs and roofs made from recycled materials) has become an increasing trend in recent years. Metal roofs made of steel, aluminium and copper often use a high percentage of recycled material. They are also lightweight. Solar roof shingles, or solar roof laminates, while expensive, decrease the cost of electricity and heating for the dwelling. Some buildings even have a living roof which consists of a waterproof membrane, a drainage layer, a growing material and plants such as sedum. This provides additional insulation, absorbs air pollution, helps to collect and process rainwater and keeps the roof surface temperature down.

Just as roofs are becoming greener, so too are the options for flooring. The use of renewable resources such as bamboo, eucalyptus and cork is becoming more common. A new version of linoleum has been developed with **biodegradable**, **organic** ingredients. Some buildings are also using floorboards and joists made from non-timber materials that can be coloured, stained or patterned.

KEY TERMS

Biodegradable

– the material will more easily break down when it is no longer needed. This breaking down process is done by micro-organisms.

Organic

– natural substance, usually extracted from plants.

Figure 3.12 Solar roof tiles provide their own solar power

Figure 3.13 A stained concrete floor can be a striking feature

DID YOU KNOW?

One of the pioneers in this type of construction is the German manufacturer HUF Haus at www.huf-haus.com

An increasing trend has been for what is known as off-site manufacture (OSM). European businesses, particularly those in Germany, have built over 100,000 houses. The entire house is manufactured in a factory and then assembled on site. Walls, floors, roofs, windows and doors with built-in electrics and plumbing all arrive on a lorry. Some manufacturers even offer completely finished dwellings, including carpets and curtains. Many of these modular buildings are designed to be far more energy efficient than traditional brick and block constructions. Many come ready fitted with heat pumps, solar panels and triple-glazed windows.

Figure 3.14 A timber-framed HUF haus is assembled off site

Architecture and design

The Code for Sustainable Homes Rating Scheme was introduced in 2007. Many local authorities have instructed their planning departments to encourage sustainable development. This begins with the work of the architect who designs the building.

Local authorities ask that architects and building designers:

DID YOU KNOW?

Local planning authorities now require that all new developments generate at least 10 per cent of their energy from renewable sources.

* ensure the land is safe for development – that if it is contaminated this is dealt with first

* ensure access to and protection for the natural environment – this supports biodiversity and tries to create open spaces for local people

* reduce the negative impact on the local environment – buildings should keep noise, air, light and water pollution down to a minimum

* conserve natural resources and cut back carbon emissions – this covers energy, materials and water

* ensure comfort and security – good access, close to public transport, safe parking and protection against flooding.

Figure 3.15 Eco developments, like this one in London, are becoming more common

Using locally managed resources

The construction industry imports nearly 6 million cubic metres of sawn wood each year. Around 80 per cent of all the softwood used in construction comes from Scandinavia or Russia. Another 15 per cent comes from the rest of Europe, or even North America. The remaining 5 per cent comes from tropical countries, and is usually sourced from sustainable forests. However there is plenty of scope to use the many millions of cubic metres of timber produced in managed forests, particularly in Scotland.

Local timber can be used for a wide variety of different construction projects:

* Softwood – including pines, firs, larch and spruce – for panels, decking, fencing and internal flooring.

* Hardwood – including oak, chestnut, ash, beech and sycamore – for a wide variety of internal joinery.

Eco-friendly, sustainable manufactured products and environmentally resourced timber

There are now many suppliers that offer sustainable building materials as a green alternative. Tiles, for example, can be made from recycled plastic bottles and stone particles.

There is a National Green Specification database of all environmentally friendly building materials. This provides a checklist where it is possible to compare specifications of environmentally friendly materials to those of traditionally manufactured products, such as bricks.

Simple changes to construction, such as using timber or ethylene-based plastics instead of PVCU window frames is a good example.

Finding locally managed resources such as timber makes sense in terms of cost and in terms of protecting the environment.

> **PRACTICAL TIP**
> www.recycledproducts.org,uk has a long list of recycled surfacing products, such as tiles, recycled wood and paving and detials of local suppliers

The Timber Trade Federation produces a Timber Certification System. This ensures that wood products are labelled to show that they are produced in sustainable forests.

Building Regulations

In terms of energy conservation, the most important UK law is the Building Regulations 2010, particularly Part L. The Building Regulations:

* list the minimum efficiency requirements

* provide guidance on compliance, the main testing methods, installation and control

* cover both new dwellings and existing dwellings.

A key part of the regulations is the Standard Assessment Procedure (SAP), which measures or estimates the energy efficiency performance of buildings.

Local planning authorities also now require that all new developments generate at least 10 per cent of their energy from renewable sources. This means that each new project has to be assessed one at a time.

Energy conservation

By law, each local authority is required to reduce carbon dioxide emissions and to encourage the conservation of energy. This means that everyone has a responsibility in some way to conserve energy:

* Clients, along with building designers, are required to include energy efficient technology in the build.

* Contractors and sub-contractors have to follow these design guidelines. They also need to play a role in conserving energy and resources when actually working on site.

* Suppliers of products are required by law to provide information on energy consumption.

In addition, new energy efficiency schemes and building regulations cover the energy performance of buildings. Each new build is required to have an Energy Performance Certificate. This rates a building's energy efficiency from A (which is very efficient) to G (which is least efficient).

Some building designers have also begun to adopt other voluntary ways of attempting to protect the environment. These include: BREEAM (Building Research Establishment Environmental Assessment Method, a voluntary measurement rating for green buildings) and the Code for Sustainable Homes (a certification of sustainability for new builds).

**energy®
saving
trust**

Figure 3.16 The Energy Saving Trust encourages builders to use less wasteful building techniques and more energy efficient construction

High, low and zero carbon

When we look at energy sources, we consider their environmental impact in terms of how much carbon dioxide they release. Accordingly, energy sources can be split into three different groups:

* high carbon – those that release a lot of carbon dioxide

* low carbon – those that release some carbon dioxide

* zero carbon – those that do not release any carbon dioxide.

Some examples of high carbon, low carbon and zero carbon energy sources are given in the tables below.

High carbon energy source	Description
Natural gas or LPG	Piped natural gas or liquid petroleum gas stored in bottles
Fuel oils	Domestic fuel oil, such as diesel
Solid fuels	Coal, coke and peat
Electricity	Generated from non-renewable sources, such as coal-fired power stations

Table 3.11

Low carbon energy source	Description
Solar thermal	Panels used to capture energy from the sun to heat water
Solid fuel	Biomass such as logs, wood chips and pellets
Hydrogen fuel cells	Converts chemical energy into electrical energy
Heat pumps	Devices that convert low temperature heat into higher temperature heat
Combined heat and power (CHP)	Generates electricity as well as heat for water and space heating
Combined cooling, heat and power (CCHP)	A variation on CHP that also provides a basic air conditioning system

Table 3.12

Zero carbon energy	Description
Electricity/wind	Uses natural wind resources to generate electrical energy
Electricity/tidal	Uses wave power to generate electrical energy
Hydroelectric	Uses the natural flow of rivers and streams to generate electrical energy
Solar photovoltaic	Uses solar cells to convert light energy from the sun into electricity

Table 3.13

It is important to try to conserve non-renewable energy so that there will be sufficient fuel for the future. The idea is that the fuel should last as long as is necessary to completely replace it with renewable sources, such as wind or solar energy.

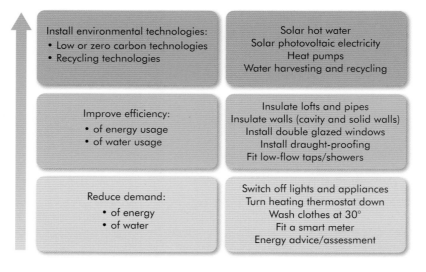

Figure 3.17 Working towards reducing carbon emissions

Alternative energy sources

There are several new ways in which we can harness the power of water, the sun and the wind to provide us with new heating sources. All of these systems are considered to be far more energy efficient than traditional heating systems, which rely on gas, oil, electricity or other fossil fuels.

Solar thermal

At the heart of this system is the solar collector, which is often referred to as a solar panel. The idea is that the collector absorbs the sun's energy, which is then converted into heat. This heat is then applied to the system's heat transfer fluid.

The system uses a differential temperature controller (DTC) that controls the system's circulating pump when solar energy is available and there is a demand for water to be heated.

In the UK, due to the lack of guaranteed solar energy, solar thermal hot water systems often have an auxiliary heat source, such as an immersion heater.

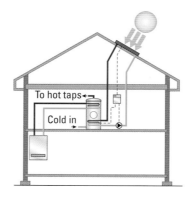

Figure 3.18 Solar thermal hot water system

Biomass (solid fuel)

Biomass stoves burn either pellets or logs. Some have integrated hoppers that transfer pellets to the burner. Biomass boilers are available for pellets, woodchips or logs. Most of them have automated systems to clean the heat exchanger surfaces. They can provide heat for domestic hot water and space heating.

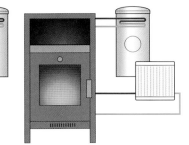

Stove providing room heat only

Stove providing room heat and domestic hot water

Stove providing room heat, domestic hot water and heating

Figure 3.19 Biomass stoves output options

Heat pumps

Heat pumps convert low temperature heat from air, ground or water sources to higher temperature heat. They can be used in ducted air or piped water **heat sink** systems.

There are different arrangements for each of the three main systems:

* Air source pumps operate at temperatures down to minus 20°C. They have units that receive incoming air through an inlet duct.

* Ground source pumps operate on **geothermal** ground heat. They use a sealed circuit collector loop, which is buried either vertically or horizontally underground.

* Water source pump systems can be used where there is a suitable water source, such as a pond or lake. Energy extracted from the water is used as heat.

KEY TERMS

Heat sink

– this is a heat exchanger that transfers heat from one source into a fluid, such as in refrigeration, air conditioning or the radiator in a car.

Geothermal

– relating to the internal heat energy of the earth.

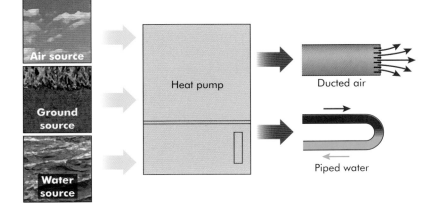

Figure 3.20 Heat pump input and output options

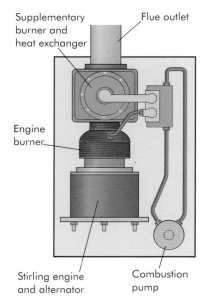

Figure 3.21 Example of a MCHP (micro combined heat and power) unit

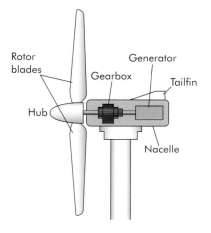

Figure 3.22 A basic horizontal axis wind turbine

The heat pump system's efficiency relies on the temperature difference between the heat source and the heat sink. Special tank hot water cylinders are part of the system, giving a large surface-to-surface contact between the heating circuit water and the stored domestic hot water.

Combined heat and power (CHP) and combined cooling heat and power (CCHP) units

These are similar to heating system boilers, but they generate electricity as well as heat for hot water or space heating (or cooling). The heart of the system is an engine or gas turbine. The gas burner provides heat to the engine when there is a demand for heat. Electricity is generated along with sufficient energy to heat water and to provide space heating.

CCHP systems also incorporate the facility to cool spaces when necessary.

Wind turbines

Freestanding or building-mounted wind turbines capture the energy from wind to generate electrical energy. The wind passes across rotor blades of a turbine, which causes the hub to turn. The hub is connected by a shaft to a gearbox. This increases the speed of rotation. A high speed shaft is then connected to a generator that produces the electricity.

Solar photovoltaic systems

A solar photovoltaic system uses solar cells to convert light energy from the sun into electricity.

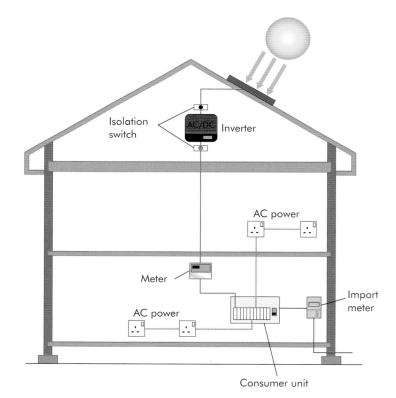

Figure 3.23 A basic solar photovoltaic system

Energy ratings

Energy rating tables are used to measure the overall efficiency of a dwelling, with rating A being the most energy efficient and rating G the least energy efficient.

Alongside this, an environmental impact rating measures the dwelling's impact in terms of how much carbon dioxide it produces. Again, rating A is the highest, showing it has the least impact on the environment, and rating G is the lowest.

A Standard Assessment Procedure (SAP) is used to place the dwelling on the energy rating table. This will take into account:

* the date of construction, the type of construction and the location

* the heating system

* insulation (including cavity wall)

* double glazing.

The ratings are used by local authorities and other groups to assess the energy efficiency of new and old housing and must be provided when houses are sold.

Preventing heat loss

Most old buildings are under-insulated and will benefit from additional insulation, which can be for ceilings, walls or floors.

The measurement of heat loss in a building is known as the U Value. It measures how well parts of the building transfer heat. Low U Values represent high levels of insulation. U Values are becoming more important as they form the basis of energy and carbon reduction standards.

By 2016 all new housing is expected to be Net Zero Carbon. This means that the building should not be contributing to climate change.

Many of the guidelines are now part of Building Regulations (Part L). They cover:

* insulation requirements

* openings, such as doors and windows

* solar heating and other heating

* ventilation and air conditioning

* space heating controls

* lighting efficiency

* air tightness.

Building design

UK homes spend £2.4bn every year just on lighting. One of the ways of tackling this cost is to use energy saving lights, but also to maximise natural lighting. For the construction industry this means:

* increased window size

* orientating window angles to make the most of sunlight – south facing windows maximise sunlight in winter and limit overheating in the summer

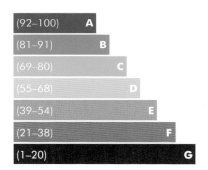

Figure 3.24 SAP energy efficiency rating table. The ranges in brackets show the percentage energy efficiency for each banding

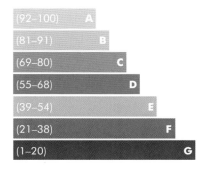

Figure 3.25 SAP environmental impact rating table

* window design – with a variety of different types of opening to allow ventilation.

Solar tubes are another way of increasing light. These are small domes on the roof, which collect sunlight and direct it through a tube (which is reflective). It is then directed through a diffuser in the ceiling to spread light into the room.

Waste water recycling
Water is a precious resource, so it is vital not to waste it. To meet the current demand for water in the UK, it is essential to reduce the amount of water used and to recycle water where possible.

The construction industry can contribute to water conservation by effective plumbing design and through the installation of water efficient appliances and fittings. These include low or dual flush WCs, and taps and fittings with flow regulators and restrictors. In addition, rainwater harvesting and waste water recycling should be incorporated into design and construction wherever possible.

Statutory legislation for water wastage and misuse
Water efficiency and conservation laws aim to help deal with the increasing demand for water. Just how this is approached will depend on the type of property:

* For new builds, the Code for Sustainable Homes and Part G of the Building Regulations set new water efficiency targets.

* For existing buildings, Part G of the Building Regulations applies to all refurbishment projects where there is a major change of use.

* For owners of non-domestic buildings, tax reduction schemes and grants are available for water efficiency projects.

In addition, the Water Supply (Water Fittings) Regulations 1999 set a series of efficiency improvements for fittings used in toilets, showers and washing machines, etc.

Reducing water wastage
There are many different ways in which water wastage can be reduced, as shown in the table below.

Method	Explanation
Flow reducing valves	Water pressure is often higher than necessary. By reducing the pressure, less water is wasted when taps are left running.
Spray taps	Fixing one of these inserts can reduce water consumption by as much as 70 per cent.
Low volume flush WC	These reduce water use from 13 litres per flush to 6 litres for a full flush and 4 litres for a reduced flush.
Maintenance of terminal fittings and float valves	Dripping taps or badly adjusted float valves can cause enormous water wastage. A dripping tap can waste 5,000 litres a year.
Promoting user awareness	Users who are on a meter will certainly see a difference if water efficiency is improved, and their energy bills will be reduced if they use less hot water.

Table 3.14

Captured and recycled water systems

There are two variations of captured and recycled water systems:

* Rainwater harvesting captures and stores rainwater for non-potable use (not for drinking).

* Greywater reuse systems capture and store waste water from baths, washbasins, showers, sinks and washing machines.

Rainwater harvesting

In this system, water is harvested usually from the roof and then distributed to a tank. Here it is filtered and then pumped into the dwelling for reuse. The recycled water is usually stored in a cistern at the top of the building.

Greywater reuse

The idea of this system is to reduce mains water consumption. The greywater is piped from points of use, such as sinks and showers, through a filter and into a storage tank. The greywater is then pumped into a cistern where it can be used for toilet flushing or for watering the garden.

Waste management

The expectation within the construction industry is increasingly that working practices conserve energy and protect the environment. Everyone can play a part in this. For example, you can contribute at home by turning off hose pipes when you have finished using water.

Simple things, such as keeping construction sites neat and orderly, can go a long way to conserving energy and protecting the environment. A good way to remember this is Sort, Set, Shine, Standardise:

Sort – sort and store items in your work area, eliminate clutter and manage deliveries.

Set – everything should have its own place and be clearly marked and easy to access. In other words, be neat!

Shine – clean your work area and you will be able to see potential problems far more easily.

Standardise – using standardised working practices you can keep organised, clean and safe.

Reducing material wastage

Reducing waste is all about good working practice. By reducing wastage disposal and recycling materials on site, you will benefit from savings on raw materials and lower transportation costs.

Let's start by looking at ways to reduce waste when buying and storing materials:

* Only order the amount of materials you actually need.

* Arrange regular deliveries so you can reduce storage and material losses.

* Think about using recycled materials, as they may be cheaper.

* Is all the packaging absolutely necessary? Can you reduce the amount of packaging?

* Reject damaged or incomplete deliveries.

* Make sure that storage areas are safe, secure and weatherproof.

* Store liquids away from drains to prevent pollution.

By planning ahead and accurately measuring and cutting materials, you will be able to reduce wastage.

Statutory legislation for waste management

By law, all construction sites should be kept in good order and clean. A vital part of this is the proper disposal of waste, which can range from low risk waste, such as metals, plastics, wood and cardboard, to hazardous waste, for example asbestos, electrical and electronic equipment and refrigerants.

Waste is anything that is thrown away because it is no longer useful or needed. However, you cannot simply discard it, as some waste can be recycled or reused while other waste will affect health or the quality of the environment.

Legislation aims not only to prevent waste from going into landfill but also to encourage people to recycle. For example, under the Environmental Protection Act (1990), the building services industry has the following duty of care with regard to waste disposal:

* All waste for disposal can only be passed over to a licensed operator.

* Waste must be stored safely and securely.

* Waste should not cause environmental pollution.

The main legislation covering the disposal of waste is outlined in the table below.

Legislation	Brief explanation
Environmental Protection Act (1990)	Defines waste and waste offences
Environmental Protection (Duty of Care) Regulations (1991)	Places the responsibility for disposal on the producer of the waste
Hazardous Waste Regulations (2005)	Defines hazardous waste and regulates the safe management of hazardous waste
Waste Electrical and Electronic Equipment (WEEE) Regulations (2006)	Requires those who produce electrical and electronic waste to pay for its collection, treatment and recovery
Waste Regulations (2011)	Introduces a system for waste carrier registration

Table 3.15

Safe methods of waste disposal

In order to dispose of waste materials legally, you must use the right method.

* **Waste transfer notes** are required for every load of waste that is passed on or accepted.

* **Licensed waste disposal** is carried out by operators of landfill sites or those that store other people's waste, treat it, carry out recycling or are involved in the final disposal of waste.

* **Waste carriers' licences** are required by any company that transports waste, not just waste contractors or skip operators. For example, electricians or plumbers that carry construction and demolition waste would need to have this licence, as would anyone involved in construction or demolition.

* **Recycling** of materials such as wood, glass, soil, paper, board or scrap metal is dealt with at materials reclamation facilities. They sort the material, which is then sent to reprocessing plants so it can be reused.

* **Specialist disposal** is used for waste such as asbestos. There are authorised asbestos disposal sites that specialise in dealing with this kind of waste.

Recycling metals

Scrap metal is divided into two different types:

* **Ferrous** scrap includes iron and steel, mainly from beams, cars and household appliances.

* **Non-ferrous** scrap is all other types of metals, including aluminium, lead, copper, zinc and nickel.

Recycling businesses will collect and store metals and then transport them to **foundries**. The operators will have a licence, permit or consent to store, handle, transport and treat the metal.

Recycling plastics

Different types of plastic are used for different things, so they will need to be recycled separately. Licensed collectors will pass on the plastics to recycling businesses that will then remould the plastics.

Recycling wood and cardboard

Building sites will often generate a wide variety of different wood waste, such as off-cuts, shavings, chippings and sawdust.

Paper and cardboard waste can be passed on to an authorised waste carrier.

Disposing of asbestos

Asbestos should only be disposed of by specialist contractors. It needs to be double wrapped in approved packaging, with a hazard sign and asbestos code information visible. You should also dispose of any contaminated PPE in this way. The standard practice is to use a

KEY TERMS

Ferrous – metals that contain iron.

Non-ferrous – metals that do not contain any iron.

Foundry – a place where metal is melted and poured into moulds.

PRACTICAL TIP

Before collection, plastics should be stored on hard, waterproof surfaces, undercover and away from water courses.

PRACTICAL TIP

Sites must only pass waste on to an authorised waste carrier, and it is important to keep records of all transfers.

red inner bag with the asbestos warning and a clear outer bag with a carriage of dangerous goods (CDG) sign.

Asbestos waste should be carried in a sealed skip or in a separate compartment to other waste. It should be transported by a registered waste carrier and disposed of at a licensed site. Documentation relating to the disposal of asbestos must be kept for three years.

Disposing of electrical and electronic equipment

The Waste Electrical and Electronic Equipment (WEEE) Regulations were first introduced in the UK in 2006. They were based on EU law – the WEEE Directive of 2003.

Normally, the costs of electrical and electronic waste collection and disposal fall on either the contractor or the client. Disposal of items such as this are part of Site Waste Management Plans, which apply to all construction projects in England worth more than £300,000.

* For equipment purchased after August 2005, it is the responsibility of the producer to collect and treat the waste.

* For equipment purchased before August 2005 that is being replaced, it is the responsibility of the supplier of the equipment to collect and dispose of the waste.

* For equipment purchased before August 2005 that is not being replaced, it is the responsibility of either the contractor or client to dispose of the waste.

Disposing of refrigerants

Refrigerators, freezer cabinets, dehumidifiers and air conditioners contain **fluorinated gases**, known as chloro-fluoro-carbons (CFCs). CFCs have been linked with damage to the Earth's **ozone layer**, so production of most CFCs ceased in 1995.

Refrigerants such as these have to be collected by a registered waste company, which will de-gas the equipment. During the de-gassing process, the coolant is removed so that it does not leak into the atmosphere.

Key benefits of using sustainable materials

In summary:

* Using locally sourced materials not only cuts down on the transportation costs but also the pollution and energy used in transporting that material. At the same time their use provides employment for local suppliers.

* In choosing sustainable materials rather than materials that have to go through complex production processes or be shipped in from other parts of the world, construction should be more efficient and have a lower general impact on the environment.

* The use of energy saving materials will have a long-term and lasting impact on the use of energy for the duration of the property's life.

KEY TERMS

Fluorinated gases

– powerful greenhouse gases that contribute to global warming.

Ozone layer

– thin layer of gas high in the Earth's atmosphere.

* Not only will the construction industry have a lower carbon footprint, but also everything they build will have been constructed using lower carbon technologies and materials.

* Protecting the local natural environment from damage by construction work or surrounding infrastructure is only part of the environmental consideration. In choosing sustainable materials to use in construction projects the natural environment is protected elsewhere, by reducing quarrying, tree-felling and the use of scarce resources

* Recycling as much construction waste as possible, particularly from demolition, means that the industry will make less contribution to landfill. Most materials except those that are toxic or hazardous can be repurposed.

TEST YOURSELF

1. What area of the construction and built environment industry would be involved in examining planning applications regarding the long-term future of an area?

   a. Town planners

   b. Surveyors

   c. Civil engineers

   d. Construction site managers

2. What is the term used to describe transport routes such as roads, motorways, bridges and railways?

   a. Services

   b. Infrastructure

   c. Commercial

   d. Utilities

3. Which of the following is an example of a corporate client?

   a. Small business owner

   b. Local authority

   c. Government department

   d. Insurance company

4. What is another term that can be used to describe a land agent?

   a. Land surveyor

   b. Quantity surveyor

   c. Estate agent

   d. Building inspector

5. Which job role involves overseeing construction work on behalf of an architect or client to represent their interests on site?

   a. Clerk of works

   b. Main contractor

   c. Sub-contractor

   d. Building control inspector

6. Which of the following is an example of a renewable energy resource?

   a. Plants

   b. Sun

   c. Wind

   d. All of these

7. What does the National Green Specification Database provide?

   a. Methods on how to recycle

   b. A list of all recycling sites

   c. A list of environmentally friendly building materials

   d. A list of components required for building jobs

8. Which part of the Building Regulations focuses on energy conservation?

   a. Part B

   b. Part G

   c. Part H

   d. Part L

9. Which of the following is an example of biomass?

   a. Coal

   b. Peat

   c. Coke

   d. Logs

10. In addition to providing heating, which of the following also provides cooling?

    a. CCHP

    b. CHP

    c. MCHP

    d. HPCP

INDEX

A

abbreviations, building industry 56
accident books 2, 9, 17
accident procedures 8–13
alternative energy sources 102–4
    biomass stoves 103
    combined cooling heat and power
        (CCHP) units 104
    geothermal ground heat 103
    heat pumps 103–4
    heat sink systems 103–4
    solar photovoltaic systems 104
    solar thermal systems 102
    wind turbines 104
alternative methods of building,
    sustainability 96–8
architecture and design, sustainability 98–9,
    105–6
asbestos 21, 24, 96
    Control of Asbestos at Work
        Regulations 3
    disposal 109–10
assembly drawings 44
assembly points 34

B

bills of quantities, costing/pricing 66
biodegradable materials, sustainability 97
biodiversity, sustainability 93
biomass stoves 103
block plans 42
brownfield sites 89–91
building materials, sustainability 99–100
Building Regulations, sustainability 100
built environment
    activities 78–9
    client types 81
    construction projects 88–92
    land types 88–91
    responsibilities 81–3
    roles 81–3
    social benefits 91–2
    work types 80

C

carbon emissions, energy conservation 101–2
carbon footprint, sustainability 94
CCHP (combined cooling heat and power)
    units 104
client types, construction industry 81
climate change, sustainability 95–6
colour coded cables, electricity 31
combined cooling heat and power (CCHP)
    units 104
combustible materials 18, 34–6
communication, good working practices 74–5
competent (individuals/organisations) 4
confidence, good working practices 75
construction industry
    activities 78–9
    client types 81
    responsibilities 81–3
    roles 81–3
    social benefits 91–2
    work types 80
construction projects
    benefits 88–92

built environment 88–92
    environmental factors 86–7
    physical factors 84–6
contaminated land 90–1
contamination 18, 19
Control of Asbestos at Work Regulations 3
Control of Substances Hazardous to Health
    Regulations (COSHH) 2, 3, 20
COSHH see Control of Substances
    Hazardous to Health Regulations
costing/pricing 61–72
    added costs 70
    bills of quantities 66
    estimates 57, 63
    hours required 70
    lead times 67
    mark-up 63
    overheads 71
    planning/scheduling 67–9
    profitability 71–2
    programmes of work 66–7
    purchasing/hiring plant and equipment
        64–5
    quotes 57, 63
    stock systems 67
    tenders 57, 63
    total estimated prices 71

D

dangerous occurrences 9
dermatitis 20, 21, 32
design and planning, roles and
    responsibilities 82
detail drawings 45
diseases 9
drawings, methods 40–2
drawings and plans 42–5
    abbreviations 56
    assembly drawings 44
    block plans 42
    detail drawings 45
    elevations 51
    floor plans 50
    hatchings 55–6
    projections 52–4
    sectional drawings 44–5
    site plans 42–4
    specification schedules 51–2
    symbols 55–6

E

ear defenders 20, 21, 32
electrical equipment, waste management
    110
electricity 28–31
    colour coded cables 31
    dangers 29–30
    Portable Appliance Testing (PAT) 28–9
    precautions 28–9
    voltages 30, 31
elevations, drawings and plans 51
emergency procedures 8–13, 34–6
energy ratings, sustainability 105
energy sources, alternative 102–4
engineering, roles and responsibilities 82–3
environment
    see also built environment
    sustainability 94
environmental factors
    construction projects 86–7
    planning process 86–7

equipment
    height, working at 26–7
    personal protective equipment (PPE)
        4, 31–3
    plant and equipment, purchasing/hiring
        64–5
    PUWER (Provision and Use of Work
        Equipment Regulations) 3–4
estimates 57, 63
estimating quantities of resources 57–64
eye protection 32

F

ferrous metals, recycling 109
finite/renewable resources 94
fire extinguishers 35–6
fire procedures 34–6
first aid 12–13
floor plans, drawings and plans 50
fluorinated gases, waste management 110
formulae, resources 57–61
foundries, recycling 109

G

Gantt charts, planning/scheduling 67–9
geothermal ground heat 103
global warming, sustainability 95
goggles 32
greenfield sites 88–9

H

hand protection 32
handling materials 22–6
HASAWA see Health and Safety at Work Act
hatchings, drawings and plans 55–6
hazards 5
    creating 17–18
    identifying 13–18
    method statements 14–15
    reporting 16–17
    risk assessments 14–15
    types 15–16
head protection 32
Health and Safety at Work Act (HASAWA) 2, 5
Health and Safety Executive (HSE) 6, 7
health risks 21
hearing protection 20, 21, 32
heat loss prevention, sustainability 105
heat pumps 103–4
heat sink systems 103–4
height, working at 4, 16, 17
    equipment 26–7
    Work at Height Regulations 4
hours required, costing/pricing 70
housekeeping 14
HSE see Health and Safety Executive
hygiene 18–21

I

improvement notices 6
information
    drawings and plans 42–5
    manufacturers' technical information
        48–9
    organisational documentation 49
    policies 47
    procedures 46–7
    programmes of work 45–6
    schedules 47–8
    specifications 47
    training and development records 49
injuries 7, 9

L

ladders 26–7
land types
brownfield sites 89–91
built environment 88–91
contaminated land 90–1
greenfield sites 88–9
reclaimed land 90
landfill, sustainability 93
lead times, costing/pricing 67
legislation
health and safety 2–8
personal protective equipment (PPE) 33
waste management 108
leptospirosis 20, 21
lifting, safe 22–3

M

major injuries 7, 10
Manual Handling Operations Regulations 4
manufacturers' technical information 48–9
mark-up, costing/pricing 63
measurements, resources 57–8
method statements, risk assessments 14–15

N

near misses 5, 10, 11
noise 20
non-ferrous metals, recycling 109

O

off-site manufacture (OSM), sustainability 98
organic substances, sustainability 97
organisational documentation 49
orthographic projections 52–3
OSM see off-site manufacture
over 7-day injuries 7
overheads, costing/pricing 71
ozone layer 110

P

PAT see Portable Appliance Testing
personal hygiene 20–1
Personal Protection at Work Regulations 4
personal protective equipment (PPE) 4, 31–3
legislation 33
physical factors
construction projects 84–6
planning process 84–6
pictorial projections 53–4
planning process

environmental factors 86–7
physical factors 84–6
planning/scheduling
costing/pricing 67–9
Gantt charts 67–9
plans see drawings and plans
plant and equipment, purchasing/hiring 64–5
Portable Appliance Testing (PAT), electricity 28–9
PPE see personal protective equipment
pricing/costing see costing/pricing
procedures 46–7
profitability, costing/pricing 71–2
programmes of work 45–6
costing/pricing 66–7
prohibition notices 6
projections
drawings and plans 52–4
orthographic projections 52–3
pictorial projections 53–4
protective clothing 32
PUWER (Provision and Use of Work Equipment Regulations) 3–4
Pythagoras' theorem 61

Q

quotes 57, 63

R

reclaimed land 90
recycling
waste management 109
waste water recycling 106–7
regulations, health and safety 2–8, 9–10
relationships, working 72–4
renewable/finite resources 94
repairing see maintenance/repair
Reporting of Injuries, Diseases and Dangerous Occurrences Regulations (RIDDOR) 2, 8, 9–10
resources
estimating quantities 57–64
finite/renewable 94
formulae 57–61
locally managed resources 99
measurements 57–8
requirements 57–64
respiratory protection 32, 33
RIDDOR see Reporting of Injuries, Diseases and Dangerous Occurrences Regulations

risk assessments 14–15
risks 5
health risks 21
roles and responsibilities
design and planning 82
engineering 82–3
surveying 82

S

safety notices 36–7
scaffold 26–7
schedules 47–8
sectional drawings 44–5
signs 36–7
site plans 42–4
social regeneration, sustainability 96
solar photovoltaic systems 104
solar thermal systems 102
specification schedules, drawings and plans 51–2
specifications 47
stock systems, costing/pricing 67
storing materials 22–6, 31
sub-contractors 6
surveying, roles and responsibilities 82
sustainability 92–111
symbols, drawings and plans 55–6

T

tenders 57, 63
toolbox talks 7, 8, 19
training and development records 49
trust, good working practices 75

W

waste control 25–6
waste management 107–10
electrical equipment 110
legislation 108
recycling 109
water, waste water recycling 106–7
welfare facilities 18–19
wind turbines 104
Work at Height Regulations 4
working platforms 26–7
working practices, good 72–5
communication 74–5
confidence 75
trust 75
working relationships 72–4

ACKNOWLEDGEMENTS

The author and the publisher would also like to thank the following for permission to reproduce material:

Images and diagrams

Alamy: Building Image: 3.14, Midland Aerial Pictures: 3.3, Peter Davey: chapter 1 opener, ZUMA Press, Inc.: 2.1; **BSA:** 2.5; **2013 © Energy Saving Trust:** 3.16; **Fotolia:** 1.1, 1.2, 1.3, 1.5, 1.6, 1.7, 1.8, 1.14, 1.15, 1.16, 3.2, 3.8; **Helfen:** 2.4; **instant art:** table 1.15; **iStockphoto:** 1.11, 2.12, 2.13, 3.1, 3.4, 3.5, 3.10, 3.12, 3.13, 3.15; **Nelson Thornes:** 1.9, 1.10, 1.12, 1.13; **Peter Brett:** 2.3, 2.6, 2.7; **Science Photo Library:** Peter Gardiner: 1.4; **Shutterstock:** chapter 2 opener, chapter 3 opener, 3.6, 3.7, 3.9, 3.11.

Every effort has been made to trace the copyright holders but if any have been inadvertently overlooked the publisher will be pleased to make the necessary arrangements at the first opportunity.